Also by Meryle Secrest

Elsa Schiaparelli: A Biography

Modigliani: A Life

Shoot the Widow

Duveen: A Life in Art

Somewhere for Me: A Biography of Richard Rodgers

Stephen Sondheim: A Life

Leonard Bernstein: A Life

Frank Lloyd Wright: A Biography

Salvador Dalí

Kenneth Clark: A Biography

Being Bernard Berenson: A Biography

Between Me and Life: A Biography of Romaine Brooks

The Mysterious
Affair at Olivetti

The Mysterious Affair at Olivetti

*IBM, the CIA, and the Cold War Conspiracy
to Shut Down Production of the World's
First Desktop Computer*

Meryle Secrest

Alfred A. Knopf
New York
2019

THIS IS A BORZOI BOOK
PUBLISHED BY ALFRED A. KNOPF

Copyright © 2019 by Meryle Secrest Beveridge

All rights reserved. Published in the United States by
Alfred A. Knopf, a division of Penguin Random
House LLC, New York, and distributed in Canada by
Random House of Canada, a division of Penguin
Random House Canada Limited, Toronto.

www.aaknopf.com

Knopf, Borzoi Books, and the colophon are registered
trademarks of Penguin Random House LLC.

Library of Congress Cataloging-in-Publication Data
Names: Secrest, Meryle, author.
Title: The mysterious affair at Olivetti : IBM, the CIA, and the Cold War
conspiracy to shut down production of the world's first desktop computer /
by Meryle Secrest.
Description: First Edition. | New York : Alfred A. Knopf, [2019] | Includes
bibliographical references and index.
Identifiers: LCCN 2018059556 (print) | LCCN 2019009701 (ebook) | ISBN
9780451493668 (ebook) | ISBN 9780451493651 (hc)
Subjects: LCSH: Ing. C. Olivetti & C. Divisione elettronica—History. |
Computer industry—Italy—History—20th century. | Computer
science—Italy—History—20th century.
Classification: LCC HD9696.2.I82 (ebook) | LCC HD9696.2.I82 S43 2019 (print)
| DDC 338.7/6213916094509048—dc23
LC record available at https://lccn.loc.gov/2018059556

*Jacket image: Olivetti Programma 101. Museo Nazionale della Scienza e della
Tecnologia Leonardo da Vinci, Milan. Wikimedia Creative Commons
Jacket design by Kelly Blair*

Manufactured in the United States of America
First Edition

For Desire, David, Philip, Matthew, Lidia, the two Annas, Annalisa, Elisa, Albertina, Francesca, Beniamino, Franco, Gregorio, Domenico, Milton, Roberto B., and all those others whose generous help made this book possible. And in memory of Roberto Olivetti.

Of truth we know nothing, for truth
lies at the bottom of a well.

—Democritus

Vespasiano da Bisticci asked Federiqo da
Montefeltro, first Duke of Urbino, what is
necessary in ruling a kingdom; the Duke
replied, *essere umano*—"to be human."

—Kenneth Clark, *Civilisation*

Contents

Preface   3

1 Oranges   5

2 "A Clear Head and a Nimble Leg"   17

3 The Convent   30

4 Enter Adriano   47

5 Giustizia e Libertà   60

6 Terror and Resolve   81

7 The Brown Affair   100

8 A Black Crossing   119

9 A Wilderness of Mirrors   136

10 Enigma Variations   155

11 The Experience of a Lifetime   178

12 High Stakes   200

13 The Curious Case of the Second Death   228

Acknowledgments   267

Notes   275

Bibliography   287

Index   291

The Mysterious
Affair at Olivetti

Preface

This book came about in a curious way. Like most authors, I had assembled a collection of short stories but had never sold any of them. One of them seemed rather better realized, so I read it carefully. It described a weekend in Washington I had spent with Roberto, the son of Adriano Olivetti, the charismatic figure who had guided the fortunes of this family-owned, Italian office equipment company during and after World War II. My good friend, Caroline Scott Despard, had fallen in love with Roberto when she was living in Italy after the war. However she broke the engagement, moved back to the U.S., and was planning to marry someone else that particular Halloween Saturday night of 1964. Roberto, who happened to be in New York, called her out of the blue. He wanted to fly to Washington and take her out to dinner. She, in a panic, called me. She was bringing her new amour, Sam, to the dinner without telling Roberto. Would I come too? I joined them rather reluctantly and spent what was left of the weekend taking Roberto around the city.

I never saw Roberto again. I discovered that in "Dinner with Roberto" I had attempted a character study of this interesting engineer, a personal portrait, because I never asked him about his company, which, as I understood eventually, was in crisis just at that moment. It suddenly became important to know whatever had happened to Roberto. I looked up his obituaries. He died in 1985 at the age of fifty-seven. The obituaries would not give the cause of death. For reasons that will become clear, I was convinced he must have died in a car accident. I persisted and finally made telephone contact with Roberto's only child, Desire, living in Italy. That illuminating conversation led to other discoveries. These included the news that Olivetti had invented the world's first desktop computer, a fact that is not reported in most computer histories. The company itself is now defunct. I found I had opened the door upon a Cold War mystery and a major industrial spy story with the Olivetti family as its victims. This account is the result.

Oranges

Carnival time in Ivrea, Italy, is celebrated each year as in many other parts of the world as a riot of exuberant excess before the sobering arrival of Lent. It is sometimes referred to as a carnival, and at others as the Festival of the Red Caps, and the Battle of the Oranges. Both are relevant because this annual ritual can be traced back for at least a thousand years. Historians believe it began as a fertility rite. Remnants of its ancient past can still be discerned; it ends, as it always has, with the burning of a tree, thought to refer to the ancient idea that if you wanted a bountiful spring, someone, or something, had to pay for it.

The custom of wearing a red cap came about during the French Revolution, in sympathy and solidarity. But when, or why, oranges became part of the proceedings no one really knows. That sunny, succulent fruit is not grown in these parts, which tend to have very cold winters—Ivrea, a town in Piedmont, is picturesquely silhouetted right up against the Alps. But some kind of battle is certainly part of the legend that has been handed down, a fable that insinuated itself into the festival in the nineteenth century and gradually took over. Time moves so slowly in Ivrea that what is measured in decades in most parts of the world is calculated in centuries here. As the story goes, a tyrannical baron in the Middle Ages demanded such a ruinous increase in taxes that his subjects were close to starving. One night the baron, exercising his *droit du seigneur,* took a miller's pretty daughter to bed. She was all prepared. At, one assumes, just the right psychological moment, she pulled out a dagger and expertly removed his head. A battle was joined. Soldiers were summoned to punish the citizens of Ivrea, only to be defeated by a handsome general who commanded a superior force. The town was saved and the pretty miller's daughter destined to be celebrated by generations of grateful citizens who have no future and for whom past glories loom large.

To visit there for a few weeks is to return with vivid impressions: green water bubbling and churning in canal locks, an empty piazza, silent in the noonday sun, shutters banging, a mist on the mountain peaks. A girl in white goes by on a bicycle, silhouetted against black foothills, there are plastic flowers on a window ledge and graffiti on the windows of an abandoned hotel. Old men amble along the river walks by day. At night teenagers smoking cigarettes huddle, chatter, and silently slip away. Ducks gather in freeform patterns against the river banks. Lanterns creak and groan and bats wheel over the chimneys. The whole town is transfixed, as if "a painted ship upon a painted ocean."

Waging war with oranges, 1950s

Once a year everything changes. Barricades appear on the many wrought-iron balconies in the main square. Merchants hammer up protective boards for their display windows. Residents begin to string up flags and bunting in the narrow alleys. Deliveries roll in; everywhere boxes are unloaded. The town is stirring, roused from its somnambulism or perhaps ennui. Elaborate floats are repainted. These will carry the groups of baron's men, who face the usual barrage of oranges that will end in the usual way, however well protected by plenty of padding and some steel helmets. There will be almost a thousand actors, descending from the hillsides and arriving from as far away as Sicily—Sicilians are in particular demand because of their accurate aim—and spectators will also be arriving in their tens of thousands. The supply of ammunition—crates of oranges laid out along the narrow alleys and lined up on the walls of the squares—is practically limitless. By the time the party is over, the cobblestone streets will be ankle deep in orange slush.

What the city of Ivrea looked like after two days of battle, 1952

A monstrous orange ball

When the ceremonies begin the floats start to arrive, with majestic slowness down the alleys, their pace dictated by their teams of horses, resplendent in crimson blankets, ribbons, and flowers, jingling and rumbling. Walls echo with confused shouts, the whinnying and snorting of the horses, running footsteps, the fluting of fifes, crashing of drums, and blare of trumpets. The windows thump and splatter with exploding fruit. As for the costumes, these echo the mad logic of ones worn by the Swiss Guards at the Vatican, all stripes and checkerboard patterns in blocks of violent red, purple, orange, blue, and green. There are fifteen teams of men, each wearing its own distinctive uniform, and a few girls as well, along with old-timers who have been battered for decades.

But the stars of the event are, first, the miller's daughter, or Mugnaia, in an ankle-length white wool dress and matching, ermine-trimmed cape with a long train. She is always accompanied by, second, her General, in all his sartorial splendor: cocked hat, black jacket, sashed, corded, epauletted, and befeathered, accompanied by white tights, black riding boots, and immaculate white gloves. They lend their benign presence to the affair, this charade, this mock victory over tyranny, that sometimes becomes an endurance test. As the hours go by, fighters drop out exhausted, squatting against the cold walls, wiping cheeks streaming with orange juice and sometimes blood. It is a sport, just a game, but every once in a while someone gets hurt. For the most part, an observer said, honor is satisfied if the soldier appears at breakfast next day with a black eye.

The red caps have become an anomaly but are always worn, as a gesture of defiance and also solidarity, conjuring up memories of a similar cap that is another part of the Piedmontese tradition. If a citizen should appear in the streets bareheaded his photograph will be published in the annual souvenir program upside-down. This faux pas is considered a serious vote of no confidence. Ivreans have not forgotten that their town was once prosperous, symbol of all that was innovative and superior in Italian design and manufacture. That was many decades ago, and now they are reduced to performing such battles for the casual delectation of the crowd. Very well; it will have to do. And so any driver who enters the city will, at some time or other, be confronted by the sculpture of a giant hand, something like a papier-mâché construction, brandishing a monstrous orange ball. No explanation is needed. Everybody knows.

The international company that brought renown and prosperity to Ivrea was Olivetti. By no coincidence, Olivetti had supported the annual affair ever since its founder, Camillo, first took part in 1880, his slim, neat figure clad in white velvet with big, puffy sleeves and a marksman's neat little hat to match. He, owner of a small but profitable electrical engineering company, was something of an artist, always coming up with machines of his own invention, and quickly saw that the quill pen was about to give way to a new machine

*A neat little figure
clad in white*

designed for the modern office. He built a primitive typewriter and then went around the town with a horse and cart, promoting its superior qualities. Quite soon he had sold enough to hire his own salesman and Italy's very first typewriter company, which by then had gone through several transformations, took off. As the company prospered, so did the annual carnival. In addition to all the paraphernalia required, the festivities included an annual buffet feast for the whole town, dances, and quantities of chocolates, candy, and bunches of mimosa that the Mugnaia throws into the crowd.

An Olivetti bride—Gertrud Kiefer Olivetti, who married Camillo's son Massimo, was the first postwar Mugnaia, well bundled up against the cold, her way lit by torches because there was, in 1947, no electricity. An Olivetti grandson, Dino's son David, or Davide Olivetti, became the most strikingly handsome General in living memory, riding, smiling, and saluting on a piebald horse in 1981. Grazia, who married the oldest of Camillo's three sons, waving and throwing candy to the crowd, became the Mugnaia six years after her marriage to Adriano, in 1956. In due course, so was their daughter Laura, always called Lalla.

Behind the blare and dazzle, the fighting, the cheers, the parades, the pomp and circumstance that followed a town's yearly assertion of

Olivetti's main building complex, set before the foothills of the Alps

victory over oppression, can be perceived the unobtrusive presence of a certain member of that family in particular, Adriano Olivetti. Along with his siblings (three sisters and two brothers), he grew up on the verdant outskirts of the town, schooled at home for several years, went on to public school and in college studied engineering. Camillo was persuaded that Adriano was ready to run the company when he was just thirty.

The son built on his father's accomplishments with even bigger successes of his own. By 1960, the company Camillo had founded with a single primitive typewriter had enlarged to factories, offices, and satellite buildings spread over fifty-four acres. Olivetti sold five kinds of typewriters, several special-purpose adding machines and calculators, teleprinters, and office furniture as well, with plants from Barcelona to Bogotá.

In an even more audacious move Olivetti had pioneered the first all-transistorized mainframe computer in 1959, running neck and neck with the one IBM, with more time and infinite resources, produced that year. Its well-built, well-conceived machines were sold in 117 countries, where it had acquired a reputation, not only for competitiveness but something more taxing: uniqueness. Their salesrooms in New York and Paris in particular reflected an elegance of design that came to be called the *Tocco Olivetti,* or Olivetti touch. A typical advertisement might "consist only of a few deft strokes from an abstract artist's pen or an intriguing geometric pattern, with the word Olivetti printed where it can be noticed but does not intrude," *Fortune* magazine reported. Or not even that. One of its early posters showed only a rose growing out of an inkwell.

The company's reach extended to its enlightened social programs. It paid better salaries than other companies. It had a shorter day and often a shorter work week. It provided a library, cinema, recreation center, first aid, a lecture program, lunches, buses, and low-cost apartments. Camillo's ironclad rule was that no worker should be fired, though he might be reassigned. Worker productivity was enviably high and so was worker loyalty. It was the pride of Ivrea. And in the fall of 1959 Olivetti had moved into the American market as well. Adriano had bought Underwood, founded in 1874, widely considered

David Olivetti, 1981, a strikingly handsome general

the model for the modern typewriter. It was the largest foreign take-over of a U.S. company that ever was. For Adriano it represented the ambition of a lifetime.

People describe Adriano Olivetti as an idealist, a futurist, a vision-ary, profoundly moral, with an uncanny ability to fulfill his dreams. To many he was considered a legend, surrounded in "an aura of superior-ity and mystery." In person he was not prepossessing. Although he had inherited his father's finely shaped head and even features, by his fifties he had grown perfectly round, what hair he had left was a wispy gray and descending chins did the rest. He often dressed with formality, looking rather like an undertaker, or perhaps on his way to a garden party at Buckingham Palace—all that was missing was a top hat. None of this seemed to matter because of the allure of his personality. Whenever he met a prospective employee, he fixed him with a dazzling regard, not quite a stare, but direct and intense. His right eye turned slightly inward, as if looking at life both outwardly and inwardly. It was odd, unsettling, yet curiously endearing.

Everyone talked about The Look. People were captivated, almost hypnotized, without knowing why. And when, instead of qualifications, bottom lines, and profit margins, Adriano talked about his dream of a better society and the virtues of community, the listener would be caught up in an idealistic fervor. One applicant said, "He hired me and I had no idea what I was supposed to do. It didn't seem important."

*Adriano Olivetti
and "The Look"*

As a rule Adriano took an active part in the three-day carnival, culminating in the Battle of Oranges, which took place on a Saturday. This particular Saturday, February 27, 1960, dawned clear and cold with not much of a wind: perfect fight weather. But Adriano could not be there for the party. He was being driven to Milan by his faithful chauffeur, Luigi Perotti, to attend an important meeting and then a lunch, followed by a weekend in Switzerland. He was making final plans for the new Olivetti-Underwood stock offering that would be offered on the Milan stock exchange the following Monday. The future plans for Underwood obsessed him. He was leaving in a week's time to inspect his prize, which would be turned into an assembly plant for Olivetti's new computers, then shipped all over the world. As usual, he had visited his favorite spa in Ischia the month before and was full of energy and well-being, his brother Dino remarked. He was fifty-eight years old.

Franco Ferrarotti, professor, sociologist, author, politician, and an old comrade, said that he, in Rome, received a long-distance call from Adriano later that afternoon. The latter was just about to board the Milan–Lausanne express.

"I was the last person to talk to him," Ferrarotti said. "He said to me, 'Franco, how are you? I'm fine. I am in a hurry. I am going to Switzerland for the weekend. Be ready to leave for Hartford,

Connecticut, on Monday, March 7." The astonished Ferrarotti wanted to know why. "Because we now have complete control of Underwood," Adriano said triumphantly. "Underwood has eighteen product lines. We will keep only three. We will use their line of distribution, which is superb. We will get the ten best engineers from Ivrea and take them to Hartford. I want you to come with us. Be ready!"

That fateful day Posy, wife of Adriano's youngest brother, Dino, saw him. "I was just walking along the street near the Hotel Principe in Milan and along came Adriano's car with Perotti driving. Perotti was honking and Adriano was waving out of the window, all smiles. I have never seen him so happy."

Ferrarotti was not, as he thought, the last person to talk to Adriano before he left on his last journey. Someone went with him to the station. He was Ottorino Beltrami, called the "Comandante," a former submarine commander who had been part of top management at Olivetti for a decade. Just why the Comandante should have been seeing Olivetti off that day is not known. What is known is that Adriano Olivetti took a seat in the back of the second-class section instead of first class, as might have been expected. We know that his mood of triumph and optimism had changed. He sat down in an empty compartment. Shortly afterward a youth (identity unknown) pulled back the compartment door and took a seat opposite him.

"The youth" as he was called, later told Swiss police that the compartment was occupied by a most "distinguished-looking gentleman," evidently Adriano in full garden party regalia. They made small talk—the train was ten minutes late leaving—and his fellow passenger picked up a magazine. He was, the youth said, "nervous, even agitated." After a while he stopped reading and was looking out of the window with unseeing eyes. "One thing struck me," the youth said. At a certain point the man's face turned a peculiar shade of purple. Then he became very pale. "I wanted to ask him if he felt unwell but he seemed so reserved I did not dare."

The train stopped at passport control at Domodossola on the Swiss border. Everyone's luggage was unloaded on the platform to go through customs control, then put back on the train, a process that took some time. While he was waiting Adriano discovered that a group of his employees, including the secretary of the company magazine, was on its way to a skiing holiday. They made plans to dine in the first-class dining car at the front of the train.

When the train stopped at Martigny, an important junction for passengers en route to Chamonix and Mont Blanc, Adriano helped the skiers get off the train. Then, as now, the roadbed between Martigny and Aigle is uneven. A passenger walking along the corridor is likely to be thrown around, particularly if the conductor is making up for lost time, as is usually the case. Accounts of the final ten minutes of Adriano Olivetti's life are as disjointed and jumpy as the ride. It seems he made uneven progress as he lurched from the front of the train, where he had been dining, to his compartment in the rear. Perhaps he was almost back in his seat when he discovered he had left his coat in the dining car. This seems to be the only conclusion to be reached, given that several passengers described seeing him making his way unsteadily along the corridor during those final ten minutes. Clearly, he was getting ready to get off the train once it reached Montreux. He never got there.

One of the passengers who saw him in the corridor was Guy Metraux, a student who gave a Paris address. He said the man was walking unsteadily, clinging to the bar running the length of the corridor on the window side. His eyes were glassy. He was stumbling. Then, he was about to fall and trying to open the door of Metraux's compartment. He collapsed, hitting his head. Metraux and other passengers rushed to help him get up and laid him down on a seat. He was semiconscious, mumbling something Metraux could not understand. A few minutes later the train pulled in at Aigle and a doctor climbed on board. It was, he said, a heart attack.

Once news reached Ivrea, the carnival's final program was canceled immediately. The funeral two or three days later was a huge affair attended by thousands, including his family, who left their hillside villas to take up their roles as principal mourners in the town. While they were out, the villas belonging to Adriano and Grazia, and that of Dino and Posy, were broken into. Nothing of value had been taken but the home offices were torn apart and papers strewn everywhere. Then it transpired that some important documents were missing. But what were they? No one was talking.

"A Clear Head and a Nimble Leg"

O ne might expect a town famous for a single object to enshrine that object repeatedly, with a chair, for example, on its coat of arms, or its promotional material, perhaps even its tax bills. But when the town fathers of Ivrea came to memorialize the name of Adriano's father, Camillo Olivetti, they chose not a typewriter, but a waterfall. Their reasons are lost to history, but the symbolism is powerful enough. A sheet of water cascades down the side of a cliff and pools beside neatly tended beds of flowers. Although this memorial is quite grand and prominently displayed at the gates of the small town it is paradoxically easy to miss. The flow of water is more of a trickle; the cliff, an indeterminate brown, is not particularly remarkable; and even the name of Camillo himself, engraved on a small bronze plaque jutting out from the waterfall itself, fades into the background. This monument to a modest benefactor is as unassuming as his lasting influence, "the moving waters at their priestlike task," as Keats wrote. He would not have wanted it otherwise.

Camillo was given the names of Samuel and David but never used them, favoring this other given name of Camillo. He was named for Camillo Benso, Count Cavour, one of the great architects of Italian unification and first prime minister of Italy; he died in 1861, three months after taking office. Camillo Olivetti was born seven years later, on August 13, 1868, second child and only son of Salvador Benedetto Olivetti and his wife, Elvira—a daughter, Rose Emma, was born in 1860. Like other Olivettis, Spanish Jews who had arrived in Ivrea in the late seventeenth century, Salvador and Elvira lived in one of the delightful villas on the hills of Monte Navale, with the foothills of

Salvador Benedetto Olivetti and his wife, Elvira Sacerdoti

the Alps clearly visible from their beautifully manicured lawns and flower beds. One sees them in old photographs. His father, balding, with a decorative edging of beard on his cheeks, sits, looking rather glum, on a tapestried chair in his best suit, with a top hat and a book on the table beside him. His mother is dressed, quietly but fashionably, in the tight-waisted, big-bustled silhouette then in fashion, almost overwhelming her petite figure. Family money on the Olivetti side came from advanced farming methods—Salvador was a noted agronomist—and the sale of property, perhaps some of it in the suburbs around Turin, where a small car company called Fiat would make its appearance.

Camillo was only a year old when his father died in 1869, far too early to have made any kind of impression on his son. As a role model, he had his grandfather Marco. His mother, Elvira, with her wide-eyed, blank stare, grew up in Modena where her father was a banker and two of her uncles were actively involved in Italy's long struggle for unification—twenty years before, they had been obliged to flee for their lives—along with Cavour. The fact that Elvira wanted her son named Camillo is significant although whether she herself took any part in the Risorgimento is unlikely. After Salvador died she moved back to Modena, where she spent most of her time. She is described as shy and diffident, as well as difficult, "perpetually confused and uncertain." "My poor mother!" Camillo wrote in 1908. "I don't think she ever understood."

A biographer of the Olivetti family notes reprovingly that Camillo was allowed to run wild and not subjected to the kind of paternal authority that would have reined in his headstrong temperament. What one generation finds flawed another will admire. For an artist the best kind of education can be no education. Camillo was, all his life, forever sketching and drawing, with machines as his favorite subject. He was an inventor in the making. He also showed unusual self-assurance. He appears in a studio portrait at the age of six or seven, posed against some fake rocks, wearing a sailor suit with matching belt, his straw hat perched like a halo on the back of his head. His classically symmetrical features, steady gaze, and confident air became the family's beau idéal. In years to come, the nicest thing you could say about a baby boy was that he looked exactly like Camillo.

Camillo Olivetti, who grew up in a house full of servants, must have soon become aware of the huge gulf separating his privileged existence from that of the average peasant in the Canavese area of Piedmont at that time. Life was, as it had been for centuries, one of unremitting hard work and poor rewards. A book about the region, published in 1960, illustrates this fact of life with surprising clarity. Clogs, the indispensable footwear for a worker in the fields, are still being made and sold at the weekly market. Women from the hillsides around arrive to do their shopping wearing heavy straw baskets on their heads. Nothing much has changed.

Peasants in the Canavese lucky enough to own their own plots of ground still expected to work it by hand, and the day lasted as long as the light did: hoeing, tilling, sowing seed, scything, harvesting—women and children beside their men.

But perhaps the worst fate was that of the casual laborer, condemned to earn his or her daily bread wherever it was to be found. One woman recalled that in 1920 there were markets for such work taking place

Camillo, aged seven

every week in conjunction with cattle markets, like the one in Saluzzo in the province of Cuneo, in Piedmont. Like cattle, they were subjected to a minute examination of their teeth, to find out if they were diseased because, as an employee, he or she would eat less. No wonder so many were deserting the countryside for the city. The fate of the rural poor became one of Camillo's lifelong concerns.

As his curiosity was aroused his grades improved. By the time he graduated in December 1891, he shot to the head of his class with the impressive score of 90 out of 100 points. He had already been accepted by an institute of fine mechanics in London for two years of advanced training. He spent part of that time touring factories, learning the practical applications of theoretical advances at first hand, and polishing his easy grasp of the English language. This mastery led to another plum assignment. Prof. Galileo Ferraris had been invited to take part in a conference on electricity being held at the Chicago World's Fair in 1893. Ferraris had very little English; Camillo could speak it. Would his prize pupil care to accompany him to the United States? Camillo would. He stayed a year.

The Chicago exhibition was timed to celebrate the discovery of America four hundred years before by Christopher Columbus, and was also called the World's Columbian Exposition. As designed by the architects Burnham and Root, as well as Frederick Law Olmsted, landscape architect, some two hundred buildings were being erected on six hundred acres, embellished by lakes, pools, and canals. It was called the White City, and from a distance its massive buildings with their cupolas, arches, and pillars seemed insubstantial, like a vision imagined by Turner. In fact it was a stage set built out of plaster

and jute, designed to be swept away as
soon as the party was over, which in
this case went from May to October
1893. The whole fair was to be floodlit
with new-fangled electricity. But by
whom? Edison General Electric bid for
the contract using the direct-current
system. Its rival, Westinghouse, used
the alternating-current system. Who
would win? The "war of the currents"
dragged on for a while. Westinghouse
kept increasing its bid until it won the
contract. The cost to the company was

Camillo as a teenager

ruinous. But it had the upper hand and the star of the show, who just
happened to be Galileo Ferraris. Camillo's first-hand experience with
American Exceptionalism had begun.

Camillo's letters from this trip have survived, and show him to
be the kind of traveler who notices everything, has something to say
about most things, and likes to write letters. His own begin after he
and Ferraris arrive in New York in August 1893. He is staying in New
York and cannot believe how simple it all is. The city has organized
avenues running north–south into a coherent grid pattern crossed by
other roads running west to east called streets. You could not get lost,
it was as easy as that. He loved American coherence and inventiveness.
He loved the efficiency and scale of it all; the Western Union offices
that were equipped with enough power to run fifty thousand telegraph
batteries. He was in awe of the Edison factory where 27,000 light
bulbs could be produced in a single day.

Shortly after Camillo Olivetti returned to Ivrea, he met three people
who were to be major influences on his life. The first was a swashbuck-
ling politician named Filippo Turati. Turati is generally considered
a leading force in the reformist movement that was sweeping Italy
on foreign and domestic policy abuses, civil rights, and the rights of
workers on such elementary issues as an eight-hour day. He was the

*Filippo Turati, a
leading reformist*

founder of a workers party that had, the year before (1893), joined forces with similarly minded groups to establish Italy's first Socialist party. That put him, and many others, in the category of conspirators against the state. Turati was usually on the move for one reason or another, such as police surveillance or threats of assassination.

Turati had trained as a lawyer but his interest in the law was almost an afterthought compared with a multitude of more pressing interests: from politics to journalism, poetry, and criminology. His rumpled, almost ferocious appearance—straggly beard, wild hair, splashy bows rather than neckties, usually askew—gave him the other-worldly charm of a man on a mission who does not have a minute to waste. On the stage he propounded, defended, exhorted, and castigated a rapt audience into submission. Camillo Olivetti was no exception. Martin Clark wrote, "[T]he chief fighters for freedom were the Radicals. The 1890s was their golden age, and their trump card was morality." The scandal of the king's harem, of colonial exploitation, the crime of child labor, and the desperate plight of the poor: all were subjects for his outrage.

When Camillo Olivetti heard Turati and how they met is unclear, but given Camillo's knack for finding and charming famous men it was probably the work of a moment. They both had wide-ranging interests and Camillo was never short for subject matter, whether the dimensions of the giant sequoias, the annual production of the Underwood typewriter company, the shortcomings of Tolstoy's *War and Peace,* or the folly of Milanese audiences—it did not matter. Turati's influence was immediate and decisive. Camillo was soon elected to Ivrea's city council as its first Socialist member. The local police took a dim view, deciding he must be "a subversive individual of socialist beliefs," and arrested him. Not that this mattered. He wrote, "I joined the Socialist Party and threw myself headlong into the struggle." He also said, "I was so carried away in those days that if only I could have found two hundred well-armed men . . . "

He was seriously thinking of starting his own business. The business climate was favorable and he had lots of wonderful ideas. The problem was that most potential workers in Ivrea and its immediate surroundings scarcely knew what electricity was, never mind how to use it.

As a student of Ferraris's, Camillo was aware that a tool to measure the voltage of direct and alternating currents was available. It was a mercury motor meter that was easy to read. But that was only the beginning; many more such instruments were needed and he was

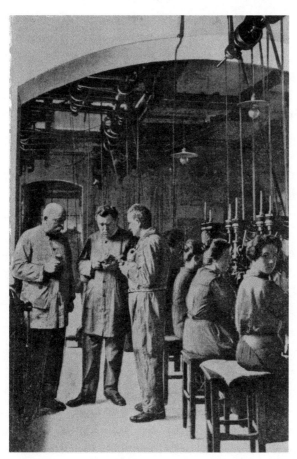

Burzio studies a problem in the early days.

probably the only man for miles who knew what they were and how to invent them. His solution was the essence of simplicity; he would start a course and hold classes in his villa in Monte Novale. He was immediately assailed with the argument that farmworkers could not learn, would not learn, lacked the necessary intelligence to master this new calling. It was probably true in some cases, but he never conceded the point. Of course they could, he said. He was lucky in that one of the first to knock on his door was a large, burly man who had trained as a blacksmith. His name was Domenico Burzio, and he turned out to be a treasure, because he was the first person to show up who knew how to melt metals.

Burzio was hired as a stoker to look after a boiler but was promoted rapidly to the role of foreman. His job was to make Camillo's designs on paper work. There was a period of experimentation during which it gradually became clear that Burzio had a natural genius for reading Camillo's mind. Or perhaps it was a talent for visualizing, and understanding the shapes and sizes of various parts needed to put a tool together. Or all of these things. It amounted to a complementary vision, the other half of the equation, one Camillo was quick to appreciate. Without Burzio, nothing was going to happen. Camillo made prominent mention of the major role this unlikely blacksmith played in his life and insisted that Burzio sit in the front row in all official photographs. It also made him realize the contributions that could come from unexpected sources, and that he must always be open to the idea that the man on the job knew more than he did. He never failed to go down the line every evening asking for suggestions for improvements.

He was finally open for business. He called his company CGS—named for centimeter, gram, and second. Within five years he was employing five hundred workers and emerging as an up-and-coming businessman.

Camillo was in his late twenties. During his period abroad he had wined and dined several attractive girls as his letters coyly suggest. A few years later he was overcome with remorse. "There came a period which, poor me, coincided with the ugliest period in my life, morally speaking. . . . May God forgive me how many evil deeds I did back then!" Whatever had caused this feeling that he had fallen into a pit of moral depravity? Could this be the same man who wrote, "I don't waste my time listening to sermons?"

Luisa Revel, the girl Camillo married, had the most important role of all to play. Like himself, she came from a persecuted sect, in this case, the Waldensian. As John Hooper points out in his book *The Italians,* one's only other option in times past was to follow the Blessed Waldo. Peter Valdo, or Waldo, was, like St. Francis of Assisi, born in the twelfth century into a wealthy family in Lyon, France, who gave away his fortune to live in poverty. His example inspired many followers. The problem was that he was required to gain the Church's permission if he wished to preach. Unlike St. Francis, who took that precaution, the Blessed Waldo did not. The Waldensians became heretics, a label they carried for centuries. Hooper notes that the Waldensians "developed an anti-ecclesiastical bias that foreshadowed some of the teachings of Luther and Calvin." They were also a few more centuries ahead of their age in their willingness to ordain women.

Down through the centuries the Waldensians were subjected to savage persecution by the Catholic Church, and by the seventeenth century they were almost wiped out as a group. Full civil and political rights did not come until 1848. Bands of fugitive followers took refuge in the Cottian Alps on the border between Italy and France, and there Luisa was born, one of fourteen children of Daniel, a pastor of the church, and his wife, Maria. The Revel family claimed to trace their lineage back to the thirteenth century. There is a photograph of Daniel in old age, seated at a table, turning the pages of a book, with a white beard, a curious plaid hat, a dark velvet jacket edged with braid, and a brace of medals on his chest. There is another photo of his tiny wife of about the same period. She is

Luisa Revel at work

Louisa, "a quiet, thoughtful girl," as a teenager (date unknown, c. 1901)

also hatted and seated, wearing the exhausted expression of a woman who has survived the birth of fourteen babies. (Many died in infancy.)

Luisa grew up in the home of a wealthy aunt whose views on a literal interpretation of the Gospels were uncompromising. Strict adherence was the rule and Luisa followed its precepts without question, graduating as a teacher from the Piccolo Institute of Divine Providence in Ivrea. She was a quiet, thoughtful girl, rather thin, who wore her hair undressed and plain cotton dresses without elaboration. This, at a moment when girls aspired to be buried in ribbons and lace, made almost immobile by vast, padded petticoats, and smothered under massive hats.

Just how they met is a matter of debate. One version is that all the local girls knew him already because of his eccentric habit of going everywhere on a bicycle, instead of the smart little carriage and pair to be expected of a well-to-do young man. This version has it that Luisa met him at her aunt's house one day when Camillo called upon one of his secretaries, who was renting a room there. On the other hand, it is said they met unexpectedly on the street. According to this version, coming upon her by chance, and struck by her beauty, he dismounted, doffed his cap, and asked her to marry him.

More to the point, she said yes and they were married in Ivrea on April 8, 1899. He was extroverted, impulsive, and never stopped talking. She was discreet, quiet, and willing to leave the starring role to him while she looked on, smiling, in the background. She had a way of persuading without contradicting, of consoling and reassuring, of accepting his confidences without judging. She was perfect for him and no doubt the other way around. Some years later, when he was traveling over the Christmas holidays, he wrote: "It is eleven in the evening and soon the New Year will begin. I want to end the year writing to, and thinking of you."

One of Camillo's pronounced characteristics was watching where the money went, from postage stamps to bus fares. Every item was carefully noted and, wherever he stayed on his American travels, finding a room for 50 cents instead of a dollar called for a triumphant letter home, along with some vague assurance to the effect that it was clean as well. Setting up his CGS company should, by his account, cost twice what he managed to scrape together—325,000 lire—and only heroic economies made it possible. Added to that was his ceaseless search for improvement that matched the architect Frank Lloyd Wright's own. As Wright said, "We aren't doing today what we did yesterday, and we won't be doing tomorrow what we did today." Not only was he artistic and creative, an American friend wrote, after receiving news of his marriage to Luisa, but he was a hard-headed businessman as well. Camillo was bound to succeed.

Quite how he fastened on building a typewriter is unclear and he must have known that such a machine, made of six thousand separate parts, made his fiddly little gadgets for measuring electricity look like child's play. Besides, it was hardly a virgin field ripe for exploitation. An Italian, Giuseppe Ravizza, a prolific as well as energetic inventor, had invented one before he died in 1885.

Camillo was also aware that Ravizza's invention had been largely dismissed by the society of his day because of its bias against technical prowess, or what he called its "anti-industrial" frame of mind. He said, "We are still the progeny of the [ancient] Romans who left all industrial work to their servants and freedmen and held it in such low esteem, that while the names of the most mediocre proconsuls, poets and actors that ever amused the decaying Roman civilization were handed down to us, the names of the supreme engineers who built roads, aqueducts and the great monuments of the Roman Empire were never divulged."

By contrast, American companies picked up Ravizza's ideas with enthusiasm, and by 1868 Remington's version, based on identical principles but with distinct improvements, was a huge success. Specifically, the QWERTY layout of the keys on the top line had been adopted and was in general use almost everywhere. Not in Italy. An Italian may have invented what Ravizza whimsically called his "clerk's clavichord," but no one wanted it.

Olivetti's first designs were in a shape that would become familiar but were limited to two rows of keys. The tattered remnants of those early experiments, now more than a century old, put one in mind irresistibly of the haphazard wooden teeth of that martyr to early dentistry, George Washington. As experiments became more polished, the race was on to improve the speed of machines so slow that the operator might as well have been setting type by hand. Improvements such as being able to make capital letters, set paragraphs, margins, and a host of other niceties appeared, but the machine was still heavy and lumbering. The goal was to build a machine that would allow an expert typist to rattle along without incurring the equivalent of a mechanical meltdown from the roadblock of too many keys bunched together. The issue of speed led to the electrical typewriter, which was faster yet, and eventually the computer keyboard, so exquisitely sensitive that strange communications appear as if by magic the moment the fingers touch the keyboard. Camillo had the opposite problem. Could a machine be invented that would respond to a lighter, faster touch?

By then Camillo Olivetti, who was in need of extra space, had moved back into the red brick building he had originally designed and built for his CGS on the outskirts of Ivrea. He brought with him a skeletal staff, and along with the indispensable Burzio set about training a new group and bought new equipment: the company's first Brown and Sharpe automatic lathes and its first milling machines. Early photographs of the interior, featureless except for some high ceilings and plenty of window light, showed numbers of young women at work wearing overalls of the ankle length then coming into fashion. Other photographs show the first machines, stacked in identical boxes stamped "OLIVETTI" being loaded onto carts, on their way to be sold, and drawn by horses. (Actual trucks came later.) Camillo, acting as his own salesman, set out with them, making the rounds of offices

*Two of the early typewriters, the M1 (manufactured
in 1911) and the M40 (1931)*

that might buy this newfangled invention. Once a machine was sold, that was not the end of it for Camillo. He would be back in a week's time, wanting to know the reaction and suggestions for improvements. Even if the client was satisfied. Camillo was never satisfied.

His first real success, the M1, was finally ready for a wider audience. "The M1 was offered to the public as a 'faster' machine thanks to a series of ideas that allowed rapid operation of the keys," Patrizia Bonifazio and Paolo Scrivano observed in *Olivetti Builds*. "For this to be achieved, the Olivetti engineers worked on the kinematic notion of the machine but they also used more sophisticated materials in the moving parts, such as forged steel, which was more elastic and longer lasting than cast iron." Getting the touch just right—this exquisitely delicate problem continued to bedevil the Olivettis for years. It was a matter of nice judgment, of fine feeling, like the final stanza of a poem. How to attain it; how to marvel in it: it was the *Tocco Olivetti* in the making.

The Convent

Quite soon after their marriage, Luisa and Camillo started having children. Since she was one of fourteen, raising a large family would have held no terrors. If she did not get lost among so many siblings, the odds are that she was seen to need more attention, for whatever reason, because she went to live with an aunt. Did she feel more wanted and needed with her own children? Was she the kind of mother who, in her turn, was all for her husband and not for her sons and daughters? It is hard to tell. What does seem clear is that Camillo had taken control of all the big decisions down to the smallest details.

Camillo had his own reasons for needing the closeness and intimacy he had missed as a boy. Far too much time had been spent at boarding school, which he resented, he did not have a close relationship with his mother, and was seldom living in the same city. The boy who never had a father now had to become one.

The first to arrive in January 1900, a year after their marriage, was Elena. Adriano appeared in rapid succession in April 1901, and Massimo at the end of February 1902. For a well-to-do family there were plenty of maids and nurses, as well as funds for frequent studio photographs of the growing family. In a clever double image, Luisa is sitting in a garden with baby Massimo, in a white dress, on her lap and Elena, standing beside her; on the right of the composition she holds another baby, Adriano, similarly attired.

Two years go by before Silvia arrives in February 1904 and another three before Laura makes her appearance in June 1907. The last baby,

Dino, is born in 1912, when Luisa is thirty-six. From the close chronology provided by a family album it is possible to chart Adriano's first months in baby dresses to the moment when he is old enough to be seen in short pants. He and Massimo could be twins, with their rosebud mouths, trim little noses, and high foreheads. The girls get equal photographic attention: Laura holding a tennis racket, Silvia in a sailor suit, and Elena with her doll. Dino, who did not resemble anybody, is seated on a chair in the middle of the group, rather than his mother's lap. Who were these children, growing up? The photographs tell us very little. In the days when studio photographers required their sitters not to move, they stand like clockwork soldiers with solemn, apprehensive, even sad faces.

While Camillo's company was in Milan they lived at No. 33, via Donizetti. But Camillo's new adventure with typewriters meant a return to Ivrea. They might still have more babies. Where were they to live? It was Camillo's decision as usual. He chose a convent.

The idea was certainly quixotic. When Camillo bought the St. Bernardino convent in 1908, it was a large, rambling, deserted building in the middle of a field connected to an old church. Nobody wanted to live there, but everyone in Ivrea knew where it was. It had been built centuries before, the fifteenth to be exact, and dedicated to a saintly figure from Siena who sounds as unappealing as he was virtuous. Several Italian Renaissance artists depicted him during his lifetime, a thin, ascetic figure whose sermons denounced the sins of sorcery, gambling, witchcraft, and much else on pain of hellfire. This was the kind of plain talking his audiences loved, particularly when he included the sin of usury. Since moneylending was often reserved for Jews, St. Bernardo was particularly known for his anti-Semitism, making Camillo's choice of homestead unexpected, like much else about his personality.

In its heyday the convent of St. Bernardino included a church containing a single nave with a cross vault, attached to buildings enclosing a cloister with, interestingly, two courtyards. What makes the church remarkable is that it is decorated with frescoes situated around and above three arched passageways leading to the presbytery. These are by a little known Piedmontese artist, Giovanni Martino Spanzotti

(1455–1528). He was a local boy who made good, having studied in Milan and married a noblewoman belonging to the Lavriano family of Chivasso, a hamlet outside Ivrea. In his lifetime Spanzotti had numerous important commissions, and one justifiably celebrated Madonna and Child on a gold background.

Just how and when he was commissioned to paint his series of pictographs on the life of Jesus is not known. They are, even at this historical distance, remarkable. His madonnas and babies show a marked departure from the stiff, formalized figures of earlier Italian religious art. Where there were once expressionless madonnas staring into space and newborns with solemn, adult faces, there are now recognizably adolescent girls holding fat, lively, sometimes wriggling babies. Spanzotti's gaze is simple and direct; as if what he had put down with his brush is happening now. Crowds stare, mutter, gawk; even a donkey looks surprised. The crucifixion, with its dark, reeling skies, agonized figures, and horrified onlookers, prefigures El Greco in intensity. Whatever caprice brought the Olivettis to the convent ended in the discovery and preservation of a national treasure that was in immediate danger of disappearing.

The nice thing about a convent is there are always plenty of sleeping cells, even if rather cold and damp. A new coal furnace provided central heating, new hardwood floors were installed, windows repaired, and at the end there was hot water for a bath. They dined by candlelight. In the meantime Camillo was in the middle of designing better, if not bigger, versions of the M1 and gaining national attention. The year his model was shown at the Universal Exhibition in Turin in 1911, he received an order for one hundred typewriters from the Ministry of the Navy. Two years after that, there was another order for fifty from the government postal system. When the firm had produced its one thousandth typewriter in 1913, every employee received a badge and Luisa was presented with a gold brooch. The future looked bright. "From that moment," Camillo Olivetti wrote later, "began the truly marvelous progress of our industry."

❧

Natalia Ginzburg, whose memoir, *Family Sayings,* is a classic in Italy, has left many vivid portraits of Camillo and his brood. Her sister, Paola, would marry Adriano. The girls came from the Levis, a cultured, close-knit family of intellectuals who had become friendly with the Olivettis and often invited them to the house.

She writes, "Adriano had many brothers and sisters, all with freckles and red hair, and perhaps my father may have liked them partly because he, too, had freckles and red hair. . . . In due course we came to know their father as well. He was a little stout man with a big white beard; in the middle of the beard he had a handsome, delicate, noble face, with shining blue eyes. He used to fiddle with his beard as he spoke, and with his waistcoat buttons. He had a small falsetto voice, rather sharp and childlike. Perhaps because of the white beard my father always referred to him as 'old Olivetti' though actually they were both about the same age. Both being Socialist, and friends of Turati, they respected and admired one another. However, when they met both always wanted to talk at the same time. . . . Old Olivetti's conversation was a mixture of the Bible, psychoanalysis and the preachings of the prophets." Her father considered that Olivetti was very intelligent, "but had very confused ideas."

Perhaps the Ginzburgs had learned about the odd way the Olivetti children were being brought up. None of the three boys was circumcised, and none of the six was baptized. They were all taught at home, which made sense, because there was no preschool available and Luisa had become a teacher. There was plenty of time for outdoor games and exercise, and everybody went to bed at nine. There was no physical punishment. The worst that could happen was the offender did not get dessert.

Natalia Ginzburg recalled that the convent "had woods and vineyards, cows and cattle sheds. Thanks to these cows they had cream cakes every day. We had a passion for cream," which their father would not allow them to eat in restaurants for fear of brucellosis, then an ever-present danger. The Olivettis owned their own cows, so the milk was safe and the Ginzburgs were forever begging to go to the convent and eat cream cakes. Their father told them to stop scrounging.

Standing like clockwork soldiers: from left, Silvia, Massimo, Luisa, Laura, Elena, and Adriano, c. 1909. Two mother's helpers stand in the background.

Camillo honestly wanted a happier life for his children than he had had for himself, so the mystery is that this did not seem to be the case. Silvia told Valerio Ochetto, Adriano's biographer, that their lives were "not joyous." Almost all of them had recurrent upset stomachs, whether from their diets, allergies, or as Ochetto believed, for psychological reasons. Adriano suffered in particular, cried, and clung to his mother. He was shy, and "acutely sensitive," Ochetto wrote.

They all understood the distinctive Piedmontese dialect spoken by their workers and servants, but Silvia was the only one who could speak it. At home, they spoke Italian. This was not enough. Camillo also decreed his children become fluent in English, a proficiency that had proved to be extremely useful in his case. When a certain Miss Ruth Philipson from Dorset was hired to join the family, her job was to teach conversational English to four of his children and live at the convent. Miss Philipson would be treated the same way as the rest of the family. "Meals are very simple, but abundant. You will have a room which is not heated except in case of illness. The corridor, however, is heated in the evening. . . . We live a very simple life, without any extravagance or luxury."

This was perfectly true. Olivetti, who still owned a bicycle, also traveled on a "motor-cycle," wearing a peaked cap, his chest plastered with newspapers to keep out the wind, Natalia Ginzburg wrote. In the days when no one owned a car, they had more than one. Camillo and Luisa felt their advantage keenly and were constantly stopping en route to offer pedestrians a lift as if in expiation. They hardly ever entertained, dressed conservatively, skied on old pairs of skis just like

*The latest advertising style for typewriters
in the early days, 1912*

everyone else, and read the Bible every night. Natalia wrote, "My mother was always saying how good and kind they were."

Some indication of what it was like to grow up in the convent was provided by Maria Luisa "Mimmina" Marxer. Her mother, Laura, died in 1934 in childbirth when she was twenty-seven years old. Camillo and Luisa took over their small grandchild and added Mimmina to the convent compound. She said, "Camillo was a very conscientious grandfather, considering that it was his duty to teach." He would summon the children, cousins and visitors included, in the afternoons for a round-robin conversation that might include politics and current events, the virtues of reading poetry, the science of crop rotation, or such minor matters as the etiquette of giving a present. This referred to the fact that whenever he went on a trip, he always returned with an armload of gifts. Mimmina recalled one occasion when he came back with an enormous pumpkin for her grandmother. Why did he do that, Mimmina wanted to know. She didn't like pumpkins. Camillo replied, "But your grandmother does. We don't give presents *we* like, we buy people what *they* like."

Her grandfather walked a lot and loved to play the piano, an upright pianoforte, until arthritis in his hands prevented it. In contrast to the rumpled tweed suits of his earlier days, Camillo took to wearing immaculate white suits, a straw hat at a jaunty angle, with plenty of room for his expanding waistline. He had no time for doctors, preferring to self-medicate with herbs, and would search for arnica flowers on his mountain walks. He had, of course, strong opinions about how to manage his rapidly growing numbers of workers, men and women. It is said that he was a hard taskmaster. He did not tolerate poor workmanship. On the other hand, he was all too aware that firing someone could lead rapidly to hardships if not destitution. So he would find work somewhere else in the factory for this person, somehow. His benevolent paternalism extended to them, as well. His door was always open, he knew everyone by name, and always found time to talk.

The same careful consideration extended to the household staff. He could not bear to see his maids down on their knees, scrubbing the floors. That demeaned them as human beings. So the new wooden floors became increasingly scuffed. It was true, Mimmina said, that he was authoritarian, even controlling. But those qualities were much tempered by the human sympathy behind them. "He was a very caring person."

In his excellent book *The Italians,* John Hooper discusses the fact that marital infidelity has always been taken more lightly in Italy than elsewhere in Europe. The death of a famous man, for instance, and subsequent obituary, will often contain a quote from his mistress as well as his wife. "Another figure who crops up quite frequently is the middle-class Italian man who drifts away from his marriage once the children have been born and embarks on a string of affairs before returning to spend his declining years with his wife." This could never be true for Camillo and Luisa still, in old age, quietly united and clearly in love.

There is a well-known episode in Adriano's young life that was dramatized in a recent quasi-documentary, *La forza di un sogno* (The Force of a Dream) shown on Italian television in 2013. It reenacts the moment when Camillo decides that Adriano, aged thirteen, should begin work on the assembly line. Adriano said later, "I soon learned to know and hate the work: a torture for the spirit. I was imprisoned for hours that never ended in the black darkness of an old workshop." For that generation, young manhood began early; in England, a boy would be apprenticed at the age of twelve. So it is logical that Camillo would do the same for his oldest son. What is less apparent is the reason why. In any event, Adriano was so seared by the experience that it took years before he could even bring himself to open a factory door.

Happily for him, he was not forced to return. Others were not so lucky. Repetitive, exhausting, unrewarding, poorly paid work: this fate swept up and deformed thousands of young lives. The architect and designer C. R. Ashbee was among those who condemned the slavery of the machine. He wrote, "Machinery untamed . . . that is the barbarism we now have to fight . . . the accursed conditions of industrial machinery in which we live." In the nineteenth century such stirrings of rebellion were widespread. Some have traced its nascent origins as far back as the French Revolution. In any event, French and British theorists were among the first to point out the dehumanizing aspects of the Industrial Revolution and advocate a more equitable distribution of wealth, better schooling, and the creation of small self-governing communities.

Others, like Elbert Hubbard, who established a colony in East Aurora, New York, believed in the mental transformation that accompanied the design and making of beautiful objects. William Morris, Ashbee, Frank Lloyd Wright, and others also advocated the return to craftsmanship, the importance of materials, and links from the beautiful to the transcendent; Wright famously referred to "sermons in stones." Italian Socialists, originally isolated groups, joined forces to agitate for better working conditions, including the eight-hour day that Robert Owen, another social reformer, had advocated, but never achieved. The desperate need for reform was in the air. If Camillo

Camillo Olivetti's first factory for typewriters, 1899

had been leading his son in the direction of Socialism by having him experience at first hand the stultifying boredom of the assembly line, he could not have hoped for a better reaction.

Adriano was growing up, a shy, sensitive adolescent with a highly developed sense of duty and moral purpose. In that respect he was his father's child. In photographs Adriano is usually found standing next to sister Elena, who became his favorite and close confidante. They still went to bed at nine o'clock, which left plenty of time for reading and Elena was reading the classics; she had recently discovered a new Italian version of *The Brothers Karamazov*. She was also reading Freud on psychoanalysis and books about astrology. Adriano, who had demonstrated a positive aversion to literature, especially novels, preferring to putter about on a carpenter's bench, relented. Before long he had discovered Rudolf Steiner, Austrian philosopher, social reformer, architect, and advocate of "spiritual science." His essay on social reform and particularly the "Threefold Social Order" so fascinated Adriano that he bought every book Steiner wrote.

Adriano was also close to Massimo, perhaps because they were only a year apart. Massimo laughed easily, loved literature and the theatre, went to concerts, and was learning the violin. Both of them had pounced on a book, *The Physiology of Love,* by Paolo Mantegazza, another original thinker who had advanced the notion that the related idea of sexuality was worthy of investigation. It was a best seller and they were in the middle of devouring this primer on sex when their mother discovered the book and threw it on the fire. Like Elena, Laura was immersed in Russian history and the rise of Communism; in solidarity, she refused to travel unless it was in second class. Silvia was secretive about who she was and how she meant to live. Although not the oldest, she was already assuming the lead. She always wanted to be the first.

It was the spring of 1918, the year Adriano celebrated his seventeenth birthday. "Dearest father," he wrote. He had decided to enlist as a war volunteer in the 4th Reggimento Alpini. He was not out to become a hero, and this was not an immature decision. Not by any means. He had given it careful thought. He wanted to do his duty as a worker and a soldier. He wrote, "You can find as many workers as you want, but I think unfortunately you will not find as many willing soldiers."

Natalia Ginzburg remembers the first time Adriano arrived at their house after he joined his Alpine regiment. "Adriano at that time had a reddish beard which was unkempt and curly, and he had long, fair reddish hair that curled down his neck." He was pale and fat—Adriano had a lifelong battle with his addiction to sugar, a running family joke. "His uniform fitted badly on his fat round shoulders and I have never seen anyone in a grey-green uniform with a pistol at his belt who looked more goofy and less soldierly than he did.

"He looked very depressed, probably because he did not at all like being a soldier. He was diffident and taciturn, and when he did speak, he spoke slowly in a very quiet voice, and said a lot of confused, obscure things while he gazed into space with his little blue eyes, at once cold and dreamy." Adriano and her brother Gino were in the same military dormitory. Gino was also serious and quiet, her father's favorite child. He would sit reading a book, and if questioned, would answer monosyllabically without looking up.

Natalia Ginzburg was one of five children. Besides Paola, there were three brothers: Alberto, Mario, and Gino. Their father, Giuseppe Levi, was a renowned anatomist and histologist. He was a professor at the University of Turin and had held other appointments at Sassari and Palermo. He was conducting important work in *in vitro* studies and the nervous system; three of his students in Turin would go on to win Nobel Prizes. He was Jewish, but an atheist; his wife, Lidia Tanzi, was Catholic and a noted hostess. Their home was a center of cultural life, frequented by intellectuals and activists as well as the occasional industrialist such as Olivetti. Surviving photographs show Giuseppe Levi as thin-faced, with a prominent nose, and smiling. But *Family Sayings* makes frequent references to his short temper, his irrational objections, and tyrannical efforts to keep his children under control.

Adriano was in love with Paola but she brushed him off. She would only talk to her brother Mario, and both were deeply resentful of their

Growing up: Adriano, Silvia, and Massimo are standing. Their mother, Luisa, is at left with Dino on her left, next to his father. Lalla and Elena are seated on the floor. Undated, c. 1920

father. Whenever he was around, they had "long faces, lifeless-looking eyes and impenetrable expressions." There would be "bad-tempered door slamming." They acted as if they were "exiled from life." Paola was growing into a beauty, which probably accounted for her father's jealous attentiveness, masked as concern. She wanted to cut her hair but he would not let her. She wanted to go to parties and dancing, and wear pretty shoes and play tennis, but no. Instead she was made to go skiing every weekend, which she hated. "However, she could ski very well. She had no style, they said, but she had a lot of stamina and great courage, and she flung herself down the slopes like a lioness." Adriano skied beautifully too, so perhaps he watched her on the slopes while she, in a rage, swept by him with her courage and disdain.

Luckily for Adriano, by the time he had finished this military training, the Armistice was declared and he never had to go to the front. That brought up the question of a vocation. To his credit Camillo strongly believed that his children should follow their own interests, also known as asking a son what he wants to do and telling him to do it. This was a bit more difficult in Adriano's case, torn between following his father and taking a totally new direction. So he compromised. He would go to the polytechnic in Turin, but not to study industrial engineering, as his father would have wished, but in the chemical industry section.

The end of World War I, which had exacted such a terrible price in millions of lives lost, mangled, and ruined, was welcomed as fervently in Ivrea as everywhere else in Europe. But one crisis gave way to others even more ominous for the future of the country. Wartime necessity had kept workers in the factories but the arrival of peace, along with the release from prison of Socialist and anarchist leaders, brought about an explosion of pent-up grievances. So many soldiers had returned home that there were not enough peacetime jobs, and unemployment reached a peak of two million in November 1919. The stock market fell and so did the value of the lira, which dropped from half of its former value to a third by 1920. While the owners of companies like Fiat, grown rich by wartime orders for munitions, prospered—the company was building a huge new factory on the edge of Turin—many of its workers were starving. Staples like bread were not to be had at any price and in August 1917 food shortages were so acute that there was a riot in Turin in which twenty-four people died.

With war's end, the unions made demands: more money, shorter hours, better job security. Fiat, which had risen from the thirtieth biggest industrial company in Italy to the third biggest, intended to put down the rioters by force. When they moved in and took over Fiat its owner, Gianni Agnelli, called for the troops, but the government declined.

Among the few factories not hit by strikes was Italy's only typewriter factory. This, as the Harvard Business School study in 1967 pointed out, was in large part due to the character of Camillo Olivetti. One morning in September 1920, Camillo met with trade unionists from the Labor Center in Turin outside the front door of his red brick factory. He invited them in for a tour. He showed them the registers that documented the treatment of his employees. Then he led them into a large assembly of his workers, and talked about his philosophy. Cooperation on work toward the same end was to be jointly rewarded, not just the company that put up the capital but the people who made it happen. His workers should share the responsibility for the company's success and therefore should share the profits. Camillo's little talks, with their references to right behavior on everyone's part, and ornamented with appropriate references to the Bible, never failed to persuade his audience. The response was, the Engineer Don Camillo knows best. Another battle won; nobody went on strike.

Like many a businessman before him, Camillo's convictions did not end at the door of his factories. He had been contributing regular articles to the Socialist daily of Turin, *The Cry of the People,* and similar dailies in Ivrea. After considering buying the Socialist paper, Camillo settled on the cheaper alternative of founding a paper in Ivrea, calling it *The Reform Action.* It ran for a little more than a year but it gave him ample space to express his views, which had to do with seizing the moment and working toward a better future, rather than a return to old attitudes and institutions. He railed against the financiers, the bureaucrats, and the crooked politicians. He argued for a new order of fraternity and justice. In this he was joined by Adriano, who was equally impassioned and impatient, writing under the pseudonym of Diogenes. He contemplated a journalist career.

Valerio Ochetto's biography of Adriano Olivetti poses the question: was he an industrialist or a revolutionary? Like his father before him, he was both, which has made them both difficult to categorize to this day. One might wonder, now that Socialist solutions have been abolished and global trade has brought the central question of employment down to its rawest solution, i.e., the cheapest labor at the lowest possible price, whether anything has changed. Was it to be continual exploitation, or could an independent, humanitarian Socialism ever be possible? It was certainly not true after World War I when cheap labor was there for the taking. What could those Olivettis possibly be agitating for? Not their own advantage, evidently. It was not rational. It was not sensible. The government should lock them up.

One of Adriano's new friends was a company engineer, Giacinto Prandi, who had joined Camillo at an early stage and became a shareholder in 1912. Adriano described him as "a noble-minded man of great intelligence and a wide and eclectic culture." They took long walks every evening for several years. Adriano had at last found a male role model in whom he could also confide, and whose advice on politics would prove to be prescient. Despite the growing power of the Communists, the labor parties, despite the marches, the protests and strikes, the Fascist party was gaining more and more adherents. Prison terms, assassinations, seizures of provincial towns, the exclusion of leftist parties from government, martial law—all these developments contributed to the foreboding Camillo and others felt when Mussolini took power in 1922. Adriano was ready to mount the barricades but the prospects for victory were not promising. When he wrote a column in praise of Gaetano Salvemini, politician, historian, and one of Mussolini's most vocal critics, Prandi took him aside. He was too young, Prandi said, to start down that road—it might have serious consequences. Adriano took the warning to heart. When Adriano graduated in the summer of 1924 he gave up what had been an ambition to become a journalist. He went to work in the factory in August

1924 at the laborer's pay of 1 lira 80 an hour. Gino Levi would join him there.

Paola Levi, 1926

&

Why did everyone think she was so happy and serene, Paola wanted to know in a letter she wrote to Adriano. She was not. She could not be more miserable. Life was like walking along a dark corridor with no light at the end; no way out. No doubt she was seriously depressed and her father's possessive control had to be one of the reasons. But even if she did marry Adriano, what difference would that make? He would be in a barracks somewhere and she would still have to live at home. It is typical of her evasiveness, her ability to assume a role, that instead of explaining why she felt trapped she spoke in vague lofty terms. She was in an "existential and sentimental" crisis. Her very soul was "disturbed and tormented." Sadness "filled her life." She spoke about the grinding sameness of her existence. She added, almost as an afterthought, she was sorry she could not marry him.

Paola was at school but did not study. This, her father felt, did not matter as she would marry anyway. What he did care about, and what Paola did not tell Adriano, was that she was in love. She had met Carlo De Benedetti, a young university friend who, Natalia wrote, "was small and delicate and had an attractive voice," who knew nothing about tissue pathology and did not ski. They went for walks together on the riverside or in the Valentino Gardens. Her father was furious because he thought to be a writer or a critic was "something contemptible, frivolous, and even doubtful." He would shout at her mother, "Don't let her go out! Forbid her to go out!" He would wake her up in the middle of the night, shouting "You have no authority!" But, as Natalia observed, their mother could never stop anyone from doing anything.

So Paola went on seeing Carlo, and her father to rage against the idea. She and Carlo were united in their adoration for Marcel Proust, who was almost unknown at the time, and in fact when Carlo became a literary critic he was the first Italian to write about Proust. The world of literature was almost unknown to the Levi family so neither parent liked the idea when Paola left the house with this young man, chatting and arm in arm. Proust became something larger in Paola's mind, not just because her parents did not like literature; he represented the forbidden future. She hung a portrait of him in her bedroom and it never left her. Perhaps Proust was, as she saw it, another martyr to his genius. Like him she was destined to spend her brilliant, doomed life in her dead parents' house, alone in a padded bedroom.

Enter Adriano

In 1920 the company brought out a new model, the M20, on which Camillo had been working for years. A great many refinements and improvements had been made, and the machine itself was handsome, with its ebony finish and gold lettering. It was a great advance on the M1—faster and almost indestructible. Camillo liked to boast that you could throw it out of a third floor window and it would not make a dent. The first M20 advertisements began to appear of pretty girls with bobbed hair, wearing flapper outfits, overlooking the new machine, their hands clasped in admiration.

Almost as soon as it arrived the M20 was shown at the Brussels International Fair and its success was assured. (In ten years, the company would sell almost ninety thousand of them.) The company was growing fast—in four years production went from four thousand to eight thousand sales a year. This rapid growth meant not just more workers, more space, and more machinery but more middle managers—a more efficient structure in effect—more specialization. In 1922 a new corporation, the Olivetti Foundry, was established, and in 1924 the Officina Meccanica Olivetti, or OMO. In a market dominated by foreign competition—in 1925, for example, Underwood was manufacturing 850 typewriters *a day*—Olivetti was barely holding its own. But in Italy its main competitor came from German imports and its superior machine, the M20, would soon dominate the national market.

At the same time Camillo and Adriano were apprehensive. Mussolini's rise brought about, along with a new authoritarianism,

Adriano Olivetti, about 1926

police powers to ferret out and assassinate enemies of the Fascist state. Salvemini, among others, was slow to appreciate the danger. He continued to publish, lecture on, and further an anti-Fascist agendum until his lawyer was beaten to death by Fascists. Salvemini found out

Giacomo Matteotti, 1924, murdered for what he knew

somehow that his name was next on the list for the Fascist death squads. He escaped, none too soon, in the summer of 1925, making his way first to France and England and finally to the U.S. A professorship was created for him at Harvard and he would lecture there until 1948.

A year before that, in the summer of 1924, Giacomo Matteotti suffered an even worse fate. Matteotti was a Milanese lawyer who became a major figure in the Socialist Party. He was elected to the Chamber of Deputies in 1919 and by 1924 was secretary general of his party at a moment when Mussolini was accelerating his terrorist attacks. Matteotti, an extremely popular, not to say fearless figure, denounced the elections that had brought Mussolini to power as rigged in the Chamber of Deputies on May 30, 1924. Two weeks later six Fascist thugs kidnapped him, bundled him into a car, and drove him twenty-three kilometers from Rome. He was stabbed to death with a carpenter's file. His corpse was discovered a few months later in a shallow grave.

Then there was the Turati affair, one that directly affected the Levis in Turin and the Olivettis in Ivrea. This admirable figure, who had worked tirelessly to better the working conditions of the poor, was now in his sixties and in poor health. (He died in 1932.) Now he was under house arrest in Milan, guarded by ten men around the clock. Something would have to be done for Turati, who could easily meet Matteotti's fate.

Natalia wrote, "In the night I heard someone coughing in the next room, the one used by Mario. . . . But this could not be Mario . . . it sounded like the cough of an old, heavily-built man."

The next morning she found a newcomer drinking tea in the dining room. He was "as huge as a bear, with a grey goatee beard. He had a very big collar size and a tie like a piece of string. He had small white hands and he was leafing through a volume of Carducci's poems." Natalia already knew who he was, but was told by her parents that his name was Paolo Ferrari, and tried to believe it. The stranger was with them for eight or ten days and whenever someone came to the door, even the milkman, he would run down the passage to his room, trying to tiptoe, "like a great shadow of a bear on the walls."

Turati had been extracted from his house arrest—a guard had been distracted long enough for him to slip out of a side door and into a car—and plans were being made to smuggle him into France. However, the winter was drawing on, the high Alpine passes were blocked with snow, and the low ones in use were heavily guarded.

The decision came to take him to Savona and then embark by boat for Corsica and freedom. On that particular night, several men in raincoats came to the house. Adriano was the only one she knew. "He was starting to lose his hair and had an almost bald-square head, surrounded by fair, curly locks. That evening his face and hair looked wind-swept. His eyes seemed alarmed, but resolute and cheerful. Two or three times in my life I saw that look in his eyes. It was the look that he had when he was helping someone to escape, when there was danger and someone to be taken to safety."

Turati arrived safely in Paris after some near disasters. But the plot was discovered and arrests were made. Somehow Adriano escaped. He hid with the Levis for several months, using the room Turati had slept in. But the Olivettis themselves were coming under suspicion for their radical views. One of Camillo's on-again, off-again magazines, *New Times,* had been selected for attack and its offices thoroughly vandalized. So it, too, had to be shut down. Adriano had to disappear for a while. In the summer of 1925, he left for England and thence by steamer from Liverpool to New York.

If, as the better part of valor, discretion dictated that Adriano Olivetti should leave Italy for a while, logic had also dictated that the only place to go was America. In 1925, when he arrived on the RMS *Caronia* (the only ship in the huge Cunard fleet named for an American), the decade was in a full-throated roar. Vast neighborhoods of small holdings were being snapped up with alacrity to make room for new money, building on a large scale, and the only place to go was up. The quintessential photograph of that era would picture a group of workers seated, having lunch on a huge supporting metal arm jutting out over a vast space in which streets were dimly visible below. For the skyscraper epitomized all that was stimulating, daring, and welcoming about the new culture; bulldozing the old in order to renew hope and vigor across the vast continent and up into the sky. Echoes of the upheaval had seeped into the European consciousness in unexpected ways, through its fashions, its dance steps ("le fox trot"), its Hollywood films—even its ability to take over such stylistic innovations as Art Deco, a French invention—and mutate them into something else. E. B. White wrote that the skyscraper, "more than any other thing, is responsible for [New York's] physical majesty. It is to the nation what the white church spire is to the village—a visible symbol of aspiration and faith, the white plume saying the way is up." Adriano was visibly impressed. The skyscrapers, he wrote, were truly grand.

By contrast, Adriano was living as cheaply as decorum would allow. He had moved out of a small hotel, paying $2.50 a night plus bath, because it was too expensive, he wrote. He settled on the YMCA on the Lower East Side, at the corner of East 23rd and Fourth Avenue, because it was cheaper: only $8 a week, all included. Adriano was counting his pennies as assiduously as his father had before him. Like his father he was on a mission to learn as much as possible that could be directly applied to improving, not just the plant, but its organization. Faster demand meant everyone must work faster. He naturally wanted to see companies directly or indirectly related to producing typewriters and office equipment. In the end he would visit over a hundred in about five months. He periodically apologized when he missed a few days of writing because everything was happening too fast.

Given his focus, Adriano's observations were more directed toward statistics, i.e., efficiency, output per man hour, cost of living, wages, and what the public was buying, than those of the usual tourist. He flicked a critical eye over New York's lack of policemen on the street, the joke of Prohibition ("Drunks are everywhere"), the general melée of vast crowds pushing and shoving rudely on the sidewalks, the extreme volatility of its markets, where values could fluctuate as much as 50 percent in a week, and even its leisure-time pursuits. On Coney Island he saw "hundreds and hundreds of people arrive in rich cars and have mad fun in idiot games." He did not think much of the movies, either.

But the real prize was Underwood, based in Hartford, the biggest typewriter manufacturer in the world. The company had introduced its Model No. 5, far in advance of the competition when it was first shown in 1900, and which sold in the millions. But then the blow fell. Not only would Royal not admit Messrs. Olivetti and Burzio, who had joined Adriano on his tour, but Underwood would not either. They were gently teased by an acquaintance who told them they had been too honest. If only they had not explained they were visiting from Ivrea. To travel across an ocean and find the door locked: for Adriano, it was like being refused admission to the Louvre. An ambition was forged at that moment that would have momentous consequences thirty years later.

By contrast the reception they received at the Ford Motor Company in Highland Park, Michigan, was almost embarrassingly warm; so cordial in fact that Adriano was emboldened to ask if he might meet Mr. Ford. (He did not.) The personally guided tour took them all around the plant, not even one of the biggest, employing something like a thousand workers and with a daily production of eight thousand cars. There were three shifts of eight hours each around the clock, paying a handsome hourly rate of $6.40. On his travels, Adriano met someone who had worked there for six months; he was well paid but had to stop because the work was so strenuous. The whole factory was a miracle of organization; not a second was wasted. "Everything runs and operates continuously. . . . I do not say that this system has no drawbacks. We witnessed a moment when the whole process came to a halt for five minutes, because one car had not been assembled correctly. That small break made for a productivity loss of five cars."

Burzio's influence was declining, and Adriano was free to launch his ideas for reorganizing the company's structure and for greater efficiency of production. The Smith Corona company had launched a very successful portable typewriter, which made all kinds of sense; Olivetti should do the same. It was Christmas of 1925 and he was in New York. It is true that America was the place for the young: bright, enterprising, full of ideas, and its technology was a modern marvel. Just the same he did not think Italy was far behind, and he confidently expected that Ivrea would catch up quickly. However, in terms of culture and civilization—not to mention politesse—there was no doubt which culture he preferred. He had expected to find lofty ideas and noble achievements. All he had discovered was the ambition to be rich. He was ready to come home. He wrote, "This childish America . . ."

—

Adriano had "walked out" with other girls and especially Ruth Philipson, the English language coach from Dorset, which in this case meant going for a drive in his car. He enjoyed her company and, it seems, saw quite a bit of her because she is mentioned in one of Camillo's letters. How would she feel? Adriano assured his father he had not made any insincere promises. He was still hoping Paola would change her mind. That had happened at last. The months he spent in hiding with the Levis may have been a factor. It is thought that she saw unsuspected strengths in him the night he drove Turati to safety. There was another incident that may have contributed to the change of heart. On one of their skiing trips Paola, in her usual reckless mood, flying down a hillside, managed to unleash a small avalanche and was temporarily buried in snow. Adriano witnessed the accident and sprang into action. His relief must have been all too evident. Perhaps they hugged each other. Adriano, that visionary man of action, was also a man of deep feeling but had enormous difficulty showing it. He would try to express himself and then feel awkward and stupid. When he hesitated, looking dreamily into space, she jumped in. She had an almost theatrical sense of timing and a special gift for having her picture taken; she would advance, turn her head slightly to one side and give the camera her most engaging smile. She was restless for new ideas, new experiences, and chance encounters. Her spontaneity bridged their gap, and there was something of a challenge, that sense of risk Adriano also enjoyed, of not knowing quite what would happen next. Surprisingly, Camillo was particularly taken with her. She was so gay, so seemingly serene, and such a beauty, with her Titian red hair, her alabaster skin, and her slim, supple figure.

It was obviously only a formality but Adriano appeared to find it necessary to ask for his father's permission to marry. By then he had returned from America and was centered in London, where he was visiting British factories. It was January 1927. Adriano listed some logical arguments for the match, such as Paola's loyalty and seriousness, along with his confidence that she would defer to him in all decisions, which Camillo seemed to think important. He added that her health "was apparently superior to mine," as if that settled it.

As might be expected, Prof. Levi was in a rage as usual. Adriano's prospects as an industrialist counted for nothing because everybody knew businesses failed every day, and why should he subject his own daughter to such a precarious future? Then there was the issue of her sullen and rebellious past. Was Adriano seriously proposing to live with such a difficult person? Adriano brushed the objections aside. His prospects were very good. (He was right about that.) As for difficult people, Elena was much worse than Paola and they had never had a single quarrel. There were family rumblings, but Adriano ignored them. The one issue he was worried about was that the good professor might not show up for the wedding. The best thing was to make it fast and simple. They were married in Turin in May 1927 in a quiet civil ceremony, followed by lunch with their respective families. Paola now had a new passport, and that meant they could both go back to London, where he had rented a house in the very upmarket area of Kensington, De Vere Gardens. Then they began their six-month honeymoon tour of Europe.

Adriano's nephew David, Dino's son, has compiled a fascinating photo album of his family, beginning with Camillo's parents and incorporating some very grand family weddings, lavishly illustrated. There is only one photo of Adriano and Paola's marriage and for symbolism it is hard to beat. To begin with, they are setting out on a journey. She has opened the back door of a car, turned and posed for a moment. She is wearing, not the spectacular wedding dress one would expect, but a simply cut summer suit in some pale color, with a curious matching cloche hat that covers her hair completely. Adriano stands behind her, wearing a dark suit and a tie, topped by a rakish beret, a hand in one pocket. They are standing close together but not linked, as if in some instinctive physical demonstration of the separateness they would feel in being together.

Besides Britain, their travels took them all over Europe, with particular emphasis on Berlin, where they stayed for some time while Adriano tried and failed to learn German. For Paola it was a release; short hair at last, and lots of beautiful, very expensive clothes. She haunted the book shops, went to museums and art galleries, saw all the latest movies. He went hunting for innovative companies, in order to study output per man-hour, statistics, desirability of materials, and what the public was buying. One of the few subjects that interested them both was the theory of personality. In Geneva they visited Charles Baudoun (1893–1963) a noted psychoanalyst and author who had made a successful synthesis of the ideas of Freud, Jung, and Adler. This led, for Adriano, to a lifelong interest in analysis although not self-analysis,

Paola holding the new arrival: Roberto, 1928

and a belief that a person's signature would tell him everything he needed to know. Returning to Ivrea they took a pretty apartment in the via Castellamonte near the factory. Paola, with her new freedom, was learning to drive. But it was also in self-preservation since Adriano was an erratic driver, liable to start the car in low gear, start talking, forget what he was doing, and continue his journey without changing gears.

Natalia wrote, "My father said nothing now, because he could not . . . forbid her or tell her to do things. All the same, he began to scold her again after a while, in fact now he scolded Adriano too. He found they spent too much, and motored too frequently between Ivrea and Turin." Such objections occasionally hit the mark. During one particularly stubborn bout of sciatica, Adriano, who had no faith in conventional medicine anyway, decided to consult a Bulgarian for something called "aerial massage." The good professor hit the roof again. "He must be a charlatan! A quack!"

His anxious forebodings increased once a baby boy arrived. He was Roberto, born on March 15, 1928. Prof. Levi was delighted, thought he was a beautiful child, and "laughed when he looked at him, because he thought he looked just like old Olivetti. 'He is exactly like the old man.' 'You would think it was old Olivetti!' my mother said, too." But that made him susceptible to the slightest setback in the baby's development, including a small temperature. "They will take him to some quack, won't they?" he would exclaim. Those ignorant parents would never get it right. The baby needed more sun or he would get rickets. As usual, the new grandmother suffered. He would wake her up in the middle of the night, shouting, "They will give him rickets. They don't put him in the sun. They should put him in the sun!"

The lessons learned from Adriano's grueling tour of a hundred American factories, along with his trips to England and Europe, had brought about a major turning point in the company's development. With the assistance of Gino Levi, now an important voice in the company, and other prominent engineers, it was decided that with better organization and a more logical system of assembly, the output could be improved by a factor of almost three. This turned out to be a shrewd assessment. Production jumped; a typewriter that once took twelve hours to complete could now be finished in four and a half hours. A new model, the M40, had been well received and the company's first portable, the MP1, was in the works, based on ideas proposed by Adriano Olivetti. Now the company was producing fifteen thousand machines a year—and all this in the Depression years. The staff had increased to nine hundred. There were thirteen branches in Italy as well as seventy-nine concessionaires. Its first foreign company, the Hispano Olivetti S.A., opened in Barcelona in 1929; a second opened in Belgium a year later. In 1933 the company celebrated its first twenty-five years and the appointment of a new managing director, Adriano Olivetti. He was thirty-two years old.

The company's rise was taking place as Mussolini, "Il Duce," and his party tightened their hold on Italian society. The Socialist Party, members of which had split off into Communist and Reformist, lost considerable power and influence, but in any event, by 1930 none of the parties could operate freely. Martin Clark observed, "There

Workers at the gate, 1930s

were . . . a few respected intellectual figures like Croce in Italy, Sforza and Salvemini in exile, keeping alive the spirit of independence and inquiry. But that was all. . . . Without institutions or organisations, the Italian anti-Fascists were like the Russian dissidents forty years before—sometimes infiltrated, usually persecuted, and always harmless." Trade unions had been organized into syndicates controlled by the state; strikes and lockouts became things of the past. The same fate awaited industrialists, who were formed into various confederations, financed and closely supervised by the state.

For Adriano Olivetti, it was "the apogee of the planning movement in Europe, planning the company's business, planning the physical plant, planning the company's urban setting, and planning the company's locale." Nowadays we know nothing is ever simple, but at the time to plan was in harmony with advanced thought and its ideals possessed Adriano for the rest of his life. "The architect was to emerge as the Planner, leader and organizer of a whole industrial society whose ends were beauty and order." In this respect he was also his father's son. Camillo's attempts to meet the urgent needs of his workers had led, by degrees, to a plan to build an ideal society. Adriano was ready to begin.

In theory there should have been nothing to create friction between Il Duce and the Olivettis since the former was, after all, just as interested in fashioning a utopia from industrial modernism. And he had been remarkably successful. In a single coup de théâtre, Mussolini drained the Pontine marshes in the Latium area southeast of Rome, a notorious source of mosquitoes and malaria, something he boasted that none of the Roman rulers had managed to accomplish. Not only did he drain the land, but he also proceeded to build five new towns on the sites, all of them on the Garden City idea. There would be thirteen altogether throughout Italy.

Garden cities were all the rage. Everyone agreed about that, so what could be wrong? Nothing, except that Camillo had been a lifelong Socialist—not a Marxist, since he advocated cooperation between owners, managers, and workers—but a potential provocateur just the same. And Adriano had written certain articles . . . Had the authorities known Adriano's role in Turati's escape, things might have been even worse. As it was, neither Olivetti had joined the Fascist Party. As yet.

In the spring of 1931 that uneasy détente came to an end. The Ministry of the Interior in Rome announced that it had come into possession of a letter written by Adriano Olivetti in December 1925 while he was in America. It was addressed to one of his mother's brothers, Ulrich Revel, who was living in Santa Barbara, California. In it Adriano made some shameful references to the Italian state as "a gang of scoundrels and murderers." He enclosed a subversive article by Gaetano Salvemini, as well as an article he had written himself, and asked his uncle if he could get it published for him in *The Nation*. This article the authorities also found offensive.

How the letter fell into Fascist hands is very curious and the six-year interval between its writing and its use by Fascist authorities is curiouser still. Adriano evidently seriously misinterpreted his uncle's sympathies. But to forward it to a repressive and authoritarian regime suggests more than disapproval; it seemed designed to do actual harm. How the family felt about all that is not recorded. It was, at the very least, embarrassing and ultimately hurtful, since it set up a police surveillance that continued for years. Although the police in Aosta took a lenient view, the Direzione Generale in Rome was threatening. While continuing to allow Adriano a passport, his movements would be closely watched "in order to better assess" his political activities.

That year, father and son joined the Fascist Party. A few months later Adriano was included in a delegation of the General Fascist Confederation of Industrialists and went to the Soviet Union to boost trade.

Giustizia e Libertà

"La famiglia è la patria del cuore." ("The family is the homeland of the heart.")

—Giuseppe Mazzini

Even though Adriano had taken over his dominant position as commander in chief of the company, at this stage Camillo was very much in charge as leader of his rapidly growing family. He was the teacher and advisor, the solver of problems, the daily manager, the admirer, critic, and to a large extent the authority in the lives of his children and grandchildren. What might seem intolerably controlling in other contexts had quite another connotation in Italy; it spelled emotional security. In photographs of the Olivettis, grandparents, parents, brothers and sisters, mothers and fathers, uncles, aunts, and cousins are closely linked physically, either as a group or in direct contact. For instance, there might be a hand on a grandfather's shoulder, an arm thrown around a wife, a mother talking to one small son while cradling another, a young couple, her arm on his back, his enclosing her waist, a grandfather with twin girls, one on each arm, or the pride of a brother escorting his sister on her wedding day. These close multigenerational patterns conceal a commonplace truth: the vital importance of the family in Italian life. Their artless gestures hold everyone in a loving embrace that shuts out the rest of the world.

Describing the "Mafia" concept in its broadest sense, Alan Friedman writes, "[S]ome sort of 'protection' had . . . always been necessary for nearly all Italians. The fragmented political landscape of Italy remains as potent a factor in the life of people today as it was before the unification of the 1860s. Even today a Milanese will feel more Milanese than Italian. Even today . . . conflicting interest groups are more important than national institutions. It is this kind of nation that places its ultimate faith in the structure of the family . . . that encourages the development of a 'Mafia' first as a familial grouping, and then as an extended family which may go well beyond blood ties to incorporate partners."

As Luigi Barzini observed in *The Italians,* "people . . . discover that all official institutions are weak and unstable in Italy: the law is flexible and unreliable, the State discredited and easily dominated . . . and society (as conceived elsewhere) has little influence. Most people tolerate small infractions, cast a blind eye on minor irritations, because to protest has consequences." He continues, "Plain speaking is often a dangerous practice. Obscurity is the rule. . . . One never knows when one's widely accepted and non-controversial opinions will turn out to be compromising and daring. One conceals one's thoughts, unnecessarily at times because, for one thing, to conceal them is never dangerous while to reveal them might be so."

To navigate in such dangerous waters calls for the kind of patience, diplomacy, and subterfuge that made *The Prince* by Machiavelli such essential reading down through the ages for any Italian who wanted to steer his modest barque into harbor. Some succeeded beyond their wildest dreams. In *Agnelli and the Network of Italian Power,* Friedman charts the rise of a middle-class Piedmontese family that turned a small automobile company, Fiat, into a multinational empire with extraordinary powers behind the scenes to direct and manipulate Italian affairs to its own advantage. As Friedman describes them, the Agnellis were "as great and powerful as the Savoys, the Medici, the Gonzagas, Sforzas, Viscontis or any other dynasty from the pages of Italian history." The Agnellis would play a pivotal role in the future of Olivetti, as a family, as a company, and particularly in the context of the model the latter presented to the world of what enlightened capitalism can achieve.

All that was to come as the young Adriano entered one of the most productive periods of his life. One sees in him a kind of resemblance to Cola di Rienzo, an extraordinary figure in fourteenth-century Italy who appeared on the scene at a moment when the popes had deserted Rome for Avignon, the city was in ruins, the poor were starving and defenseless, and the rich were barricaded inside palaces that looked more like armed camps. Cola, a handsome and intelligent young man, grew up in the country, developing an interest in antiquity and legend. He was a brilliant orator. Adriano was not—he proselytized in print—but their reactions to the misery they found around them were the same. Adriano's attributes closely mirrored those of Cola: "literary, artistic, vague and contradictory ideas, practically unrelated to the contemporary world . . . the dream of building 'a new State,' inspired by ancient history in which peace, law and virtue would prevail; a genuine love for his people, his country and their glorious past . . . and a desire to avenge his people's ruin and humiliation."

Another astute observer of the Italian scene, John Hooper, has pointed out that, to no one's surprise, businesses that start in an Italian family stay in the family, from the mom-and-pop corner store to the august heights of the Agnellis. Ferrero, the makers of Nutella, is still owned by the son of the founder. "Silvio Berlusconi's business empire will likewise continue to be a family affair; his daughter by his first marriage runs the holding company, Fininvest, while her brother is the deputy chairman of Berlusconi's television group. . . . Most of the great Italian fashion empires also grew up around families: the Benettons, the Ferragamos, the Guccis, the Versaces." Camillo took up the services of his dead daughter Laura's husband, Carlo Lizier; brought his daughter-in-law Paola's brother Gino Levi into his orbit; made use of Elena's husband, Arrigo Olivetti (no relation), and other relatives and descendants. It goes without saying that Massimo, who went on to play a leading role, began at the beginning just as Adriano had done. Dino's talents would be put to their first-class use for the Olivettis' American division. But as Hooper also notes, "[F]amily firms have a poor record of investment in research and development, which is becoming increasingly vital for businesses," an observation that would be very much to the point for the Olivettis in the postwar years.

At this stage the issue was not so much research and development—everyone knew that, to keep ahead of the market, continual refinements would be needed—but how much to invest in bricks and mortar. Camillo, who had done his fair share of adding on to the original 1908 building, was against doing any more. His native caution told him that, not only was Europe in the midst of a depression, but malign new forces were at work. In Spain, Francisco Franco, an authoritarian general, would seize power after a bloody civil war. Olivetti had business interests there, Belgium as well, and who knew when some other distant relative's spiteful behavior would cause further mischief with the Fascist regime. He and Adriano had joined the party in the admittedly slim hope of heading off another disaster. Hitler had seized power and it would become absolute after the death of Hindenburg in 1934. The German dictator's secret plan, a "final solution" to the Jewish question, had been leaked as early as 1931 and Jews were already leaving Germany. Was it wise to be investing any more money just then? Camillo wrote to a friend, "[T]he world's horizon is so fraught with threats that it is best not to risk expanding long-term businesses."

Had he known what kind of building Adriano wanted to put in place, Camillo would probably have liked the idea even less. In art a new wave of Futurists, in love with progress and technology, was celebrating the arrival of the train, the airplane, and the motorcar in drawings that attempted, symbolically or otherwise, to convey the forward thrust of a new age. A famous Cubist painting of 1912 by Marcel Duchamp, *Nude Descending a Staircase,* had done just that, causing a sensation when it was exhibited at the New York Armory show some twenty years before. Futurism, Fauvism, Constructivism, Abstract Expressionism, Dadaism—such developments were giving rise to a parallel revolution in architecture as men like Charles Edouard Jeanneret, a Swiss-French architect otherwise known as Le Corbusier, Walter Gropius, Adolf Meyer, and Ludwig Mies van der Rohe ushered in the new era. The machine had arrived and turned the old values upside down. Now what counted was the extent to which an architect could successfully adapt his creation to the requirements of the new materials and mass production, how well he could build for the new spirit of progress, revolution, industrialism, and social betterment. Just how this was to be done was the mantra of the Bauhaus, a German art school founded by Walter Gropius just after World War I that became one of the most influential advocates of modern architectural design.

The new buildings, whether houses, apartments, or commercial structures, shared the qualities of the groundbreaking Bauhaus complex, that is to say they were uncompromisingly boxy, streamlined, uniform, regimented, looking like the factories in which they had probably been assembled and certainly like "machines for living," that phrase of Le Corbusier's that was to haunt him ever afterward. Ornamentation, whimsical shifts of direction, unexpected nooks and crannies—all these, being evidence of the bad old days, were abolished. The new architects espoused the doctrine of simplification, purification, the nobility of glass-curtain walls and transparent volumes, and dedicated themselves to a technological and functional future in which fancy and feeling played no part. The Villa Savoie, which Le Corbusier built in 1928, is a white box on stilts surrounded by a horizontal band of windows in a wooded lot; rooms look out onto treetops and down into the gardens. Hugh Honour and John Fleming wrote, "[T]he cubic volume of the Villa Savoie is hollowed out on three sides, mainly on the south-east and south-west so that sunlight can flood right into the centre of the building. And since load-bearing walls had been eliminated the interior could be left quite free. . . . That the building would be impossible to comprehend from any single viewpoint only confirms how completely Le Corbusier had achieved his objective of total interpretation of outer and interior space." In America the architect Philip Johnson, one of the most influential advocates for the International Style (as it was called) took the idea of an invisible division between inside and outside to its logical extreme. The house he built for himself in New Canaan, Connecticut, had no exterior walls at all—they were all glass—and none inside either, except for the bathroom. The house was the subject of a famous joke by Frank Lloyd Wright, who went to see it. "Ah yes!" he said to the architect. "Philip Johnson! You're the man who builds those little houses and leaves them out in the rain." Other observers, such as the critic Lewis Mumford, would also come to criticize the "antiseptic cleanliness and irreducible bareness" of the new style. Mumford wrote in 1964 that Mies van der Rohe "used the facilities offered by steel and glass to create some elegant monuments of nothingness."

If the Victorian darkness and squalor of the original Olivetti building had repelled and depressed Adriano Olivetti, one can readily see why he would be attracted to clean, boxy lines of the Bauhaus-inspired alternative. He had made contact with a team of young architects, Luigi Figini and Gino Pollini, who had recently graduated from Milan Polytechnic and opened a studio there in 1929. They had joined the "Gruppo 7" with other young architects fired with enthusiasm over the example set by Le Corbusier, and Figini had built his own "rationalist" house modeled closely on the Villa Savoie; no doubt Adriano went to see it. He and his partner were engaged to build a new factory wing lined up along the avenue and connecting directly with earlier

*Le Corbusier with a model of his famous
house, the Villa Savoie, 1930s*

buildings. "The innovation introduced in the first expansion of the plant was based on models similar to architectural developments taking place in the United States and elsewhere in Europe; the creation of large ribbon windows and the use of reinforced concrete for the supporting structure in order to form large spans in the work area, were two of the novelties." The spirit of the Bauhaus, as epitomized by Le Corbusier, would arrive in Ivrea even if, for prudent reasons, the new wing would not be built upon stilts.

The new wing signified much more than a need for more space. One suspects a son's natural desire to make his mark, to distinguish the company from its many competitors. What he was obviously looking for was a way to align technical achievements with aesthetic ones at the highest possible level; to be, as Jean Cocteau used to admonish, ahead of the avant-garde instead of trailing along behind it. He would design the most functional and beautiful machines, his publicity and graphic arts would operate at the most discriminating levels, his stores would be miracles of high style, and his offices and factories would grace the landscape instead of defacing it. Adriano's taste in art is never mentioned but, from his immediate acceptance of Bauhaus diktats, the conclusion is that he was attracted to the boldness of the designs, the simple but powerful solutions to complex problems and the satisfying logic of its interiors, even if others found them severe. The larger, overwhelming issue for him was that the name of Olivetti should gain lustre because it was directly linked to the ultimate in design and execution.

No sooner had the plans for the new building taken shape than Adriano Olivetti set his sights on Le Corbusier. His idea was that the farmland surrounding the convent be developed, over time, into a kind of Olivetti village. The plan would be supervised by the company, make use of Figini and Pollini, and engage Le Corbusier for its overall concept. Sensing trouble, Le Corbusier was not interested. It was possible, he said politely, that a random group could accomplish something like this, but (he did not say) pigs could fly. On the other hand, if he were given the proper authority, he would be delighted and for a relatively modest cost.

Other ideas came and went. In the end, the Olivetti village was never graced with the name of Le Corbusier. Adriano had to be content with whatever aura of the great man's reputation could be detected in the work of his pupils and admirers. For instance, the next redesign of the Olivetti typewriter, Studio 42, which arrived in 1936, was made by Alexander Schawinsky, an alumnus of the Bauhaus.

Adriano had prudently postponed work on the new addition until his father left for a long holiday. When Camillo returned and saw what was happening, Roberto Olivetti, Adriano's son, recalled, he was not happy. He gave voice in later years to the conclusion that it was a mistake to promote his sons, Massimo included, so young. In fact, he had made just the right choices. Only he could have made the start, but only they, with their particular perspectives, could have guided it from a local venture to the global heights it would soon achieve.

ℰ℘

Shortly after they were married, Adriano and Paola, who had been living in Ivrea, moved to the Piazza Castello in Milan and there Lidia was born in 1932, a fair-skinned, blue-eyed, red-headed baby like her brother. Adriano needed to be in Ivrea where everything was happening, but Paola craved the fun and mental stimulus of a large city; now that she had her own car she was always on the move. Adriano

Lidia, aged five or six, 1933   *Roberto, aged six or seven, 1935*

had acquired a chauffeur so that he could come and go at will, but the trip was considerable so he probably stayed in Ivrea quite often. Perhaps this is when he began his custom of early morning appointments beginning at 6:30 a.m., after breakfast of course. When he did arrive home it was late, and how much sleep he ever got is another question.

Glimpses of their married life are recorded by Natalia Ginzburg and show that, even if not actively engaged himself, Adriano's circle consisted of men and women more or less working behind the scenes to depose Mussolini. As part of the anti-Fascist underground Giuseppe Levi kept them apprised of the latest developments, usually bad. Their famous friend Vittorio Foà, a prominent Socialist politician and trade unionist, was in prison. So was Riccardo Bauer, editor of the anti-Fascist *Non Mollare.* Ernesto Rossi, economist and journalist, who was involved in the anti-Fascist Giustizia e Libertà operating out of Turin, had been arrested in 1930 and would remain there for the next thirteen years. Natalia wrote, "My father thought of them with a mixture of veneration and despair, since he didn't believe they would ever be released."

Adriano was calm and confident. He knew from "one of my sources" that Fascism was about to self-destruct. Yet, as Natalia observed, one began to suspect that his source was, probably, one of the palmists he visited in every city he traveled through. "He said some were very good and had accurately described his own past, and some even 'read minds.' Adriano considered thought-reading a fairly common ability." Her mother loved to have him visit because he always arrived with fresh predictions about the wonderful future in store for the Levi family. This soon included another seminal figure, that of Leone Ginzburg, publisher, writer, teacher, and dedicated anti-Fascist, who arrived in their circle and fell in love with Natalia. They were married in 1938 on the eve of war.

Adriano was the one who brought news to the family about Mario's narrow escape from prison. Mario, now an economist, was working for Olivetti in Ivrea and usually came home to Turin on the weekends. Unknown to his parents, he was secretly supporting

Giustizia e Libertà. He had a girlfriend living in Switzerland and would sometimes go there on weekends instead, bringing back with him stacks of anti-Fascist pamphlets, well hidden, in a car being driven by a friend, Sion Segre, another anti-Fascist. They had done it so often they grew careless and, this particular weekend, did not even try to conceal their cargo.

As luck would have it this was the weekend when, as they were crossing from Switzerland into Italy at Ponte Tresa, Italian customs officials, who were looking for smuggled cigarettes, found pamphlets instead. The police made them get out and were marching them along the riverbank dividing the two countries when Mario, fully dressed in a heavy overcoat, jumped into the river and made for the Swiss bank on the opposite side. He came within a hairsbreadth of being shot by a Swiss guard. Luckily, the man held his fire and Mario, in cold, raging water, weighed down by his coat, was rescued just in time.

Natalia wrote, "Adriano's face wore that expression of mingled happiness and fear in the time of danger that it had at the time of Turati's flight. He put a car and a driver at my mother's disposal but she did not know what to do, or where to go. She kept putting her hands together, saying with a mixture of happiness, admiration and alarm, 'In the water, in his overcoat!' "

Having never had a father himself, Camillo treated his sons as he imagined a father would: that is to say, affectionate and concerned, but uncompromising, expecting the high standards from them that he demanded from himself. The situation was further complicated by an evolving rivalry between the brothers, especially Adriano and Massimo, who were in a permanent tug-of-war having to do with which of them should be in charge.

Where they were all in agreement was their duty to provide generous welfare programs. In 1926 Camillo had built six workers houses near the factory. They were a rather delightful anachronism in their nondescript surroundings, looking as if they had strayed from an Alpine village: pitched roofs, plastered walls, symmetrical windows and doors, and a big enough garden for a vegetable plot. This was mandatory because the ever-practical Camillo thought his workers needed a way to keep their families fed if they were ever out of work.

Nothing could be in greater contrast than the first row houses and apartments designed by Figini and Pollini in the decade before World War II. Their stripped-down interiors looked like the low-cost solutions that they actually were: just walls, with a predictable picture window at the far end. They were at least in beautiful surroundings. These minimalist designs would take on rather more inventive forms because workers and managers housing continued to be built postwar. But for the moment, the most one can say is that they provided a roof over the rain and it did not take long to get to work. Many ambitious plans were in the making for this "Olivetti village" on what had been Camillo Olivetti's farmland. And schools were high on the list. It was becoming a genuine community.

The nascent planner, social reformer, and philosopher in Adriano's complex personality were hardly satisfied with these issues, which, after all, were easily solved with buildings and enough money. He was already ruminating over larger philosophical questions. What was the meaning of life? How could a man's surroundings bring about a harmonious inner state of mind? How much would the rational design of cities lead to the appearance of a sane society? Why not start with Ivrea?

Thanks to Pollini, Adriano had made contact with another important group of architects, the BBPR (Gian Luigi Banfi, Ludovico Barbiano di Belgiojoso, Enrico Peressutti, and Ernesto Nathan Rogers), who were passionate proponents of Le Corbusier's dicta. In *Invented Edens,* Robert Kargon and Arthur Molella explained the major challenge such ideas presented for architects, industrialists, and politicians. They needed to join forces, not just to plan cities but whole regions, in the interests of a "functional city."

In the opinion of the reformists, European cities were uniformly decrepit, their decaying centers centuries old and rife with poverty, disease, and crime. Tearing down such monuments to social ills would be an elementary beginning. Since they had grown like Topsy in a higgledy-piggledy fashion down through the ages, clearly they also needed to be radically remodeled. Massive clearance would make way for a bigger scale: bigger houses, bigger lots, wider roads, and many more cars. Logic would triumph, once the whole chaotic mess had been sorted out, into clearly defined areas for home, work, leisure, and communication. Adriano thought all this made perfect sense. Figini and Pollini were put to reorganize Ivrea's particular muddle. New roads would be built once the old eyesores had been removed. Ivrea needed a specific industrial district. It needed harmony, logic, and order. The result would be splendid. Adriano was so convinced that he devoted years to the plan's development and paid for some very elaborate models.

What he never considered was whether Ivrea needed a makeover. True, it was not an architectural masterpiece, so it would not matter if some of the buildings came down. This particular belief ran counter to the fact that Ivrea, a modest town in the foothills of the Alps, is remarkably coherent architecturally. It presents a consistently serene aspect to the world, with its arched stone bridges, its uniform roofs of red-clay tile, its wrought iron balconies, faded green shutters, and bell towers. Ochre, its prevailing wall color, is restful to the eye, its

Workers' housing of the 1930s

town hall is a gem, and its squares, with their colonnades and outdoor cafés, invite a mood of relaxation and reflection. Happily, the radical transformation never took place. Such development as there is occurs on the fringes and a ring road keeps the traffic well separated from the pedestrian center.

On the other hand the imprint of Modernism—the attempt to insert the rational ideal in the context of a harmonious whole—has been jarring wherever it has been applied. The planners seem to have taken no account of those quaintly narrow alleys, shifting perspectives, and all the spontaneous arrivals of squares and green spaces that go to make up the evolution of a town over the centuries. The work of the past was an anachronism in an age devoted to logic and the impersonal reordering of priorities.

A prime example of this kind of architectural "reform" can be seen at the Villa Casana on the outskirts of town, a large country house standing on its own grounds. Such an establishment was first erected in the sixteenth century. The present house was embellished and enlarged in 1818 by the architect Giuseppe Maria Talucchi for Count Alessandro Vallesa in an elegant pink brick, edged with stone facings around windows and down the corners of walls; there are even Romeo-and-Juliet balconies. Before being used for company offices and exhibition space, it served as a nursery school.

There never was a third floor but there is one now, adding a suite of offices on top of the roof. These consist of a horizontal band of picture windows fitted with uniform, vertical black metal blinds, the whole topped by a thick block of concrete. A more jarring juxtaposition would be hard to find. An even more egregious example, the La Serra complex, graces the center of Ivrea and will be discussed in a later chapter.

Curiously enough the Villa Belli Boschi, one of the few homes Adriano ever owned, was left relatively untouched. This delightful, wisteria-covered country villa in the prevailing colors of ochre and green, its shuttered windows looking out over gardens, trees, and fields, has retained its languid charms over the centuries. One of the upstairs rooms, originally a bedroom, now the Ivrea office of the Fondazione Adriano Olivetti in Rome, has retained Adriano's suite of

Adriano's office, still intact to this day, and the view of
the foothills of the Alps from his window

furnishings as they were when he left them in 1960. These turn out to
be Art Deco in feeling, the coffee tables, chairs, and bookcases in the
same fruitwood, with softly rounded corners. Their "streamlined" style
suggests they were probably acquired when the vogue first appeared in
1937. Adriano's desk still has its position of honor in front of a large
picture window that surveys the foothills of the Alps. One imagines
him there, looking out into the distance.

Adriano Olivetti's ambitious plan for Ivrea was never implemented.
At the time, he had taken up the challenge of bringing Le Corbusier's
vision into the impoverished mountain valleys to the northwest of
Ivrea, the Val d'Aosta. He was working in collaboration with tourist
boards and was quick to win over local financial interests. Just the
same, to develop the town of Aosta and its surroundings was ten times
more ambitious than rerouting roads through Ivrea. As presented,
it was not so much a master plan as a succession of four separate
plans, as Patricia Bonifazio and Paolo Scrivano write in *Olivetti Builds.*

These were: the development of the mountain face of Mont Blanc, a separate plan to build settlements in the Breuil valley at the foot of the Matterhorn, a third, for a ski resort in Pila, and, finally, a new urban plan for the city of Aosta itself. These plans should have taken account of the topography of the landscape and local construction traditions, as Bonifazio and Scrivana also note, i.e., the context. Instead, they were rigid geometric designs imposed on the landscape. The work of Figini, Pollini, and the BBPR these plans were launched at the Galleria della Confederazione Nationale Artisti Professionisti in Rome in 1937. The government was approached to send a representative to the opening of the exhibition. Mussolini looked over the plans—several tables full of models, aerial photographs, and masses of statistics. He took out a blue pencil and, to his credit, wrote "No." He never went to visit Adriano's new buildings in Ivrea either. So much for development on the slopes of Mont Blanc.

Adriano's innovations at that period were extraordinary. The first planer milling machine, designed by his father, was produced in 1938. The portable typewriter, which would become de rigueur for generations of expatriate writers, was catching on. Studies for adding and calculating machines led to their production in the 1930s. The first Italian teletype, the T1, which Massimo, who had inherited his father's singular talent, invented, went into production in 1935. A new building on the via Jervis was proving tricky to connect up with other buildings. Figini and Pollini, who were continuing to experiment, added rows of slender concrete columns that soared skyward and ended with a flourish. They looked for all the world like lilypads and were a direct steal from similar models created by Frank Lloyd Wright in 1936.

Adriano's days must have been so full of plans, budgets, overdrafts, and other challenges such as the psychology of workspaces and load-bearing metals, that how much of a companion he was to a wife who would rather talk about art, literature, and the cinema, is a moot point. She might have been interested in the magazine he had just founded but then again, perhaps not, since it was concerned with industrial organization. He seemed to be a man so possessed by the myriad issues he faced every day that he was oblivious to everything else. Including his wife.

Along with the Levis, the Olivettis continued to affect a normal, everyday reality; they could hardly afford to do anything else. But they were aware that the Levi family home in Turin was a meeting place, a salon for a small but important group of anti-Fascists who were actively planning the overthrow of Mussolini in the years before World War II. In *Modern Italy,* Martin Clark writes that, at that time, anti-Fascism was limited to dissident groups of intellectuals who might produce "a few" clandestine publications, then disappear into years of exile. Giustizia e Libertà, founded in 1929 by a group of Italian emigrés in Paris—Carlo and Nello Rosselli, Ernesto Rossi, Carlo Levi (no relation), and others—was different. The group not only managed to include adherents from a spectrum of political opinions but accomplished publicity stunts like dropping anti-Fascist leaflets from airplanes. They even made plans to bomb Mussolini's villa in Rome on the Piazza Venezia, but nothing came of it. Giustizia e Libertà "had considerable appeal for young intellectuals, and soon it had an underground organization in Italy—by 1933 it may have had as many adherents as the Communist Party." That same year, in Germany, the Nazis turned violent; Jews were kidnapped and tortured, paraded through the streets, and sixty thousand people went to prison. Another ten thousand fled, arriving as refugees all over Europe and America.

Similar attempts to menace and intimidate were happening in Italy, if not to quite the same extent. Anyone suspected of sedition was imprisoned or exiled and that, in the case of Mario Levi's daring escape, led directly to his family and friends. Sion Segre, who had been caught with him smuggling leaflets, went to prison. Mario's father, Giuseppe, went to prison and so did Giuseppe's son Gino. Leone Ginzburg shared the same fate and so did Carlo Levi. Natalia Ginzburg wrote, "[T]hen one day an article appeared in the newspaper, headlined 'Group of anti-Fascists discovered in Turin ganging up with Paris exiles.' 'Ganging up,' my mother repeated in distress. This phrase sounded full of dark threats. She wept in the sitting room with her friends round her." Prof. Levi was released after two weeks but

*Portrait of Paola
by Carlo Levi*

Gino served for three months. Perhaps it was after that experience that Gino Levi became Gino Martinoli, making him marginally less easy to identify. He became one of Olivetti's three directors and successfully ushered it through the difficult war years. As for Paola, although she seemed calm she was scared. She decided that the main entrance to their building was "under observation," because someone in a raincoat was always hanging around. She was sure she was being followed.

This could well have been true. Among the well-known dissidents who were frequenting her family's salon in those days was a flamboyant figure destined to play a major role in her life. He was Carlo Levi, born in Turin, also to a well-to-do Jewish family, who had trained as a doctor. An early supporter of Giustizia e Libertà, he, along with Leone Ginzburg, drafted its manifesto. He decided he wanted to devote his life to art rather than medicine and applied himself diligently to his task. He favored portraits in loose, pale washes and with indeterminate delineations meant to capture a mood, but which barely resembled their sitters, including many that he painted of Paola. As for more general subjects, these appear stiff and labored and reaching for a symbolic synthesis that eludes them.

By contrast Carlo Levi was a born writer, with a gift for clear, poetic prose that paints a vivid mental image. He wrote many books but the account for which he is famous, *Christ Stopped at Eboli*, refers to a period in the 1930s when he was sent into solitary exile to Lucania, the Italian version of Siberia. This, now called Basilicata, is an impoverished area in southern Italy where periodic wars down through the centuries, malaria, bandit raids, and indifference by central governments had left its inhabitants uneducated, badly fed, and sometimes starving. Their plight was so desperate, as Levi wrote, that its inhabitants felt even Christ had abandoned them. They had been bypassed "by Christianity, by morality, by history itself." Observing a gravedigger at work, Levi writes, "The ground was littered with calcified bleached bone, flowering out of the graves and worn away by the wind and sun. To the old man these bones, the dead, animals and spirits were all familiar things, bound up, as indeed they were to everyone in these parts, with simple everyday life. 'The village is built of the bones of the dead,' he said to me . . . twisting the toothless hole that served him for a mouth into what might have been meant for a smile." Levi's evocative and haunting descriptions became an overnight sensation when the book was published in 1945. It was praised for its humanism and also his gift for bringing national attention to what had been an ignored and forgotten people.

Carlo Levi's affair with Paola Olivetti seems to have begun shortly after the birth of Lidia in 1932. Paola took some kind of job as secretary for his artistic career, working in his studio and no doubt tried to help him conceal his activities on behalf of Giustizia e Libertà. It was too late. He was constantly being hauled in for questioning and incarceration for varying periods of time. Once released he would return to Paris where he had established residency, but as soon as he was back in Turin he would be pounced upon once more. The longest period he served, in Lucania, came as a result of the Mario incident and lasted for a year (1935–1936). He spent it at work at his easel but was also pressed into service as a doctor since there were very few in the area. His arrivals were as sporadic as his affections because there were at least three other women who came and went as well: Vita Gourevitch, a beautiful Latvian girl; Maria Marchesini, who would have married him; and Anna Maria Ichino, who already had a child by another man. Paola was in love with him too. She wrote to tell him how much she needed him. "Only you, I wish you were close to me. I see a great light . . . a halo of gold dust around your head, as in an old painting."

While Paola was hurtling down her particular emotional ski slope, dislodging who knows what avalanche of feelings, her lover was buried

Carlo Levi, left, with a friend, Renato Guttuso, in Venice, 1936

in a world so alien that, he confided to Paola, he could hardly exist. He hated the primitive peasants with their arcane rituals, so remote from everything, in his eyes, that made life worth living. She was sympathetic: "I'd be dead already if, like you, I was buried for . . . months in a group of houses away from the world, among the veiled women, the goats and witches and angels." She wanted him to change the subject. "Speak to me about love, not about Aliano." As for Adriano he must have known she had left him in spirit, if not in person. Perhaps, after trying so hard to win her, he had given up. He may have had other liaisons himself. It would have been characteristic of him to immerse himself in his work, which was an even more immediate crisis, not to mention the growing evidence of a coming European war. Events were moving fast in and outside Italy, and his first priority had to be protecting the company.

The latest concerns were the signs that Mussolini, in the thrall of Hitler, was moving closer to punitive laws aimed at Jews. Something bad was going to happen. A long press campaign had started in 1938 in support of a supposedly scientific report, "The Manifesto of Race," asserting the superiority of Europeans over other races. Martin Clark wrote, "Many business firms closed down; Jewish children were expelled from state schools; one in twelve university teachers . . . lost their jobs; the brilliant physicist Enrico Fermi left Italy in protest." There were a few exceptions. Camillo Olivetti was deemed to have provided essential service to the state and was excluded. Because they had been baptized in the Waldensian faith, Adriano and Elena were considered Protestants, and also excluded. But their brothers and sisters—Silvia, Massimo, Dino, and Laura—were not.

In 1936 Carlo Levi was given a premature pardon and released.

He returned to Turin but was under constant surveillance. Paola and her two children were also in Turin, for a time at least. Giuseppe Levi had lost his chairmanship at the university and had taken a position in Liège, Belgium; his wife joined him there. Natalia and Leone had just married but now they were both in prison. All their friends were leaving or planning to go; when Carlo was offered a trip to America in 1937, he took it. He still had an apartment in Paris and when he returned, Paola joined him there. She and Adriano had agreed to separate (divorce in those days was impossible), and the legal papers came through in 1938. By then, Paola had given birth to Carlo's baby.

Little Anna was born in March of the same year. A photograph of her, about six months old, in her mother's arms, is included in the Olivetti family album. She was first called Annetta, after Carlo's mother, but the name was soon dropped. As for Paola, she stands in a garden wearing a pretty summer dress, her hair ruffled, looking both triumphant and defiant, as if flaunting the arrival of the small person who would forever bind Carlo to her. But the immediate issue was how to give Anna some legal protection. To be illegitimate, with a convicted Jew for a father and a Jewish mother, was unfortunate. Adriano came forward with a solution. He was legally a Protestant. So he would claim paternity. The requisite forms were filed and Anna Levi became Anna Allegro Olivetti.

One fine spring day in April 2016 I spent two hours as guest of Anna Olivetti and her husband, Antonello Nuzzo, in their studio apartment tucked away in a quiet corner of Florence. Anna is now an architect, as is her husband, and the walls were covered with photographs, drawings, portraits, and paintings, as well as shelves of books. Anna, trim and short, was wearing a black leather jacket and talking about her childhood as she pointed out the many paintings of her mother by Carlo Levi, as well as some nude studies. She said that the large, unframed canvases were always kept in a secret, locked room, because her mother by then was living with Mario Tobino, an Italian poet, writer, and psychiatrist, who was very jealous. Then Anna asked, with a half smile, "You know my father is Carlo Levi? You didn't know?" She went on to say she did not know herself until five or seven years ago because "my mother always kept that fact very, very secret, but not only my mother. Everybody. For instance, my sister Lidia, who is about nine years older than me, when I was little, almost my mother, never told me anything even when my mother died." (In 1986.) "It was really strange for me."

She was asked how she felt at the news. She gave a laugh of dismissal. "I was not upset. The reality for me is, my mother was a mother and a father to me."

As it seemed, the child who might have cemented the relationship between Paola and Carlo appeared to have destroyed it. Carlo could barely manage to keep himself out of jail—once war began, he dared not go back to Italy—and naturally did not want a wife and baby to share the same fate. There was a further factor, his mother's violent reaction to the news. Paola wrote, "I never thought I would have days as bad as this. Perhaps my life will become easier. Because here I am [she was in Turin], reduced to a family situation that is socially untenable. Anna is my sole consolation. She becomes every day more intelligent and lovable. She understands everything, even complicated things. Her gestures are full of grace and she is so good and cheerful." It is clear Paola felt rejected and was defenseless against Carlo's mother's accusations. "I regret very much that you did not talk to your mother before leaving, and that was a big mistake. I do not know what she thinks now, last night she did not speak to me personally, as others were there. Soon I will go to Ivrea, and who knows what will happen. You will always think ill of me and this saddens me greatly. Unfortunately everyone thinks badly of me, and I do not know what to do."

Two days later she wrote again. "You should write your mother a letter saying I am not such a bad person. She wants to believe that I raped you, I betrayed my husband, that I am a terrible whore. Goodbye, Carlo dear, it was so nice to live with you, and I am sorry that you also had to pay dearly. I cannot wait to see you again. I am like a tree without soil or leaves. Barely alive. I want you and I love you."

Terror and Resolve

A month after the outbreak of World War II, in October 1939 Camillo was writing to a Spanish colleague that things at the factory were going well but "the whole world is very nervous because, no one knows where it will end. Most of the population is hoping that [we] Italians do not have to be dragged into the war." Having reached his seventies Camillo was bothered by arthritis, usually taking self-prescribed cures after which he declared himself much improved. He still spent hours every morning drawing ideas for new machines that, as yet, only existed in his imagination. He had taken up a new cause, that of Unitarianism—he would joke that he was probably its only Italian adherent. He was offering his usual unsolicited medical advice to anyone who would listen, and explaining artlessly that he should have been a doctor. As for events in and outside the factory, not to mention the well-being of his distant sons, daughter, and grand-children, it would be safe to assume he demanded to know what was happening, minute by minute.

The Convent, at that moment, contained its usual assortment of children, grandchildren, relatives, and the odd guest. Elena and her husband, Arrigo, were living there, along with their children, and so was Massimo's family, his wife, Gertrud, her mother, her son by a previous marriage, and three children of their own, all girls. There were also seven servants, meaning more than twenty adults and children to feed, care, and plan for. Silvia had passed her medical exams with flying colors and emigrated to Argentina to escape the racial laws, where she was making new friends. Dino served in the Italian campaign in

Camillo, 1940

Ethiopia doing convoy work. He then went to Boston and enrolled at the Massachusetts Institute of Technology where he completed his degree in mechanical engineering. There he met Rosamond "Posy" Castle, a winsome beauty from Quincy, Massachusetts, married her, and moved to Brazil where he was working in the Olivetti plant.

On the assumption that Paola and the children would be safer in the outskirts of Florence, in Fiesole, Adriano bought a villa for them and moved them into it. La Piazzola, although not large, was situated

Paola can be faintly seen to the right of the oversized flower pot in the courtyard
of La Piazzola, Fiesole, where she, Roberto, and Lidia spent the war years.
The photograph is undated, but is likely to have been taken in 1943.

in perfect juxtaposition with several famous villas on one of the most
picturesque hillsides of that enchanting village. The area has been
much favored since the eighteenth century by British expatriates, not
to mention later arrivals such as Bernard Berenson, with his much-
admired Villa i Tatti, and Gertrude Stein, the American author, and
her friend, Alice B. Toklas. Directly above La Piazzola is the Villa
Medici, named for Giovanni, son of Cosimo Medici the elder, who
built it in the fifteenth century as a social center for aesthetes and the
high intelligentsia. It owes its fame to Lorenzo the Magnificent, who
inherited the property in 1469, and to a series of elegantly formal
gardens terraced below it on a steep slope that wrapped around the
hillside.

The house, with its commanding position, ideal proportions, and
stately rooms, did indeed become a magnet for artists, philosophers,
and men of letters, Berenson included. In the eighteenth century it
was bought by Lady Orford, sister-in-law of Horace Walpole, celebrated Englishman of letters. By the time Paola arrived in Fiesole,

the Villa Medici belonged to Lady Sybil Cutting, mother of the writer Iris Origo. It had been more or less annexed by Berenson's coterie after she married Geoffrey Scott, an English writer and landscape architect, who was a more or less permanent fixture in the i Tatti circle. Scott and his close friend, Cecil Pinsent, also a landscape architect, added to the villa's réclame by creating its third terrace, laid out in Italian style along the hillside. Scott is best known for his study, *The Architecture of Humanism,* and also his biography of Madame de Charrière, the eighteenth-century Dutch novelist of *The Portrait of Zélide,* which is dedicated to Sybil.

The steeply terraced hillside descends to La Piazzola, which, although it has only two bedrooms, has a very large reception room and enviable views from its terraces. Adriano doubtless knew from his usually reliable "sources" that Bernard Berenson, along with others in Fiesole's Anglo-American colony, were under the personal protection of Count Gian Galeazzo Ciano, Mussolini's son-in-law and foreign minister, who had not approved of Italy's entry into the war with Germany and made his displeasure known when it did. After that it was clear to those aristocrats and members of the high intelligentsia who remained in their palatial villas that they had nothing to fear even if they were Jews. Such protection, it turned out, was destined to end. Meanwhile Berenson, that elegant and accomplished art historian, authority on the Italian Renaissance, authenticator, dealer, and bibliophile, went on buying, selling, and entertaining nonstop in his somewhat austere villa on the hillsides of Florence.

As Adriano left Florence to return to Ivrea, one cannot know what he was thinking, but the odds are good that he was already far beyond his present predicament as a jilted husband in a world gone mad. What mattered was getting America into the war, as Adriano knew it would, and Germany's defeat was guaranteed. What would soon be needed was a blueprint for a better, saner society. What would a new Arcadia be like? Curiously enough Berenson, who kept a diary until he died, periodically mulled over this idea. He liked the idea of Socialism. "In a sense, there would be no rewards. The pooled products of everybody's labor would be for everybody to use freely. As there would be no prestige connected with possessions . . . people would want only what they could enjoy." So he mused as he sat and pondered on the hills of Fiesole.

If only . . . Adriano Olivetti was, by contrast, a practical political thinker and moralist. He imagined the future as the means to an altruistic end, that of freeing mankind from ignorance, superstition, brutality, and injustice. Like Madame de Staël he was an indignant theorist, rebel, and

Giuseppe Pero

reformer. He believed in the perfectability of mankind and was ready to face any test with that mingled expression of terror and resolve that Natalia Ginzburg so eloquently described. However, the first objective was to keep the family safe and the factory functioning. He was beginning to lean more and more on Massimo, who turned out to have inherited his father's engineering creativity. Massimo's first big patent, the teletype machine, which was a huge success, would lead to other inventions. He had taken a degree in Turin and had made the requisite tours of the U.S., U.K., and Germany before taking up an influential post in Ivrea. He met and married a divorced aristocrat, Gertrud Kiefer, while he was touring in Germany. To her three children, Massimo and Gertrude added three more, all girls.

As expected, Adriano seemed unaffected and went on working as if there was no war and no danger to anyone. He was still largely involved in building projects and was turning his attention to the new studies of work environments that had been carried out in the U.S. in the 1930s. These focused on psychological influences, such as how a building might be designed, how space should flow, how colors influenced mood, and the importance of the right lighting, all of them aspects that interested Adriano in the cause of bringing about the optimal environment for creative thought. He was also moving decisively to modernize the company's corporate image, i.e., the way it presented itself to the world. He was assembling an extraordinary team of designers who were breaking new ground in the art of attracting attention by means of posters, leaflets, booklets, advertising, and show windows.

One of the newcomers was "Xanty" Schawinsky, who, after working at the Bauhaus for four years, had joined Olivetti. "The stern austerity of the Bauhaus was somewhat softened and made more poetic by the Italian atmosphere, but here was the same formidable use of photography, the brilliant play of type, and the disdain for irrelevant frippery that had marked the Bauhaus style."

Giovanni Pintori, who would put his distinctive stamp on Olivetti's graphic images, was hired when he was twenty-four and served as the company's masterful promoter for the next thirty years. One of his best-known posters was filled edge to edge with a chaotic jumble of numbers; the Olivetti trademark was dead center, promoting both Olivetti's technological achievements and its ability to bring order out of chaos. Adriano's imprimatur had to be sought in every detail, whether it was the color scheme used or the shape of a space bar.

Pintori said, "He not only chose the men who did the work but he gave suggestions and criticisms of the finished work down to the last detail. Particularly with the architects, he was constantly involved in the job—but with us, too. He was a man teeming with intuition. He had a style, and although he might not be able to fabricate it in any special way, he knew how to communicate its sense and then judge its results."

Or, as his son Roberto wrote, "His organization was informal. There were no regular meetings. Whenever an important decision was to be made he called in the people in whom he had faith. He surrounded himself with very intelligent people and listened to their judgment. . . . He was constantly involved in helping the younger executives to prepare for future positions. The top two levels of the organization were in constant contact with him. Everyone knew his ideas and goals. He had a strong, forceful personality which helped him to sell the others on his goals.

"He never really understood or cared for the accounting system. . . .

He left the finance and control work to [his chief financial officer] Dr. Giuseppe Pero," who was stumpy and white-haired and had been the company's director-general since 1930.

Then, one gathers just for fun, Adriano decided to found a publishing house. He went to Luciano Foà, a major literary agent in Milan, in the fall of 1941 to discuss his ideas. Foà was enthusiastic. The two men began to make plans but these were cut short when the British began bombing Milan. So offices were moved to Ivrea, and the NEI, or New Editions Ivrea, symbolized by a dove with an olive branch, opened its doors. As usual Olivetti surrounded himself with an energetic young staff, including Giorgio Fua and Erich Linder and began to buy up rights: Hemingway, Keynes, Jung, Freud, and Kierkegaard in translation. Meanwhile the factory was still making typewriters. In 1942 there were more than four thousand employees and 64,000 sales. Very soon afterward Olivetti, with its knowledge of metalwork, was building guns, machine guns in particular, for Mussolini's war.

☙

Posy Castle was a tall, dark-haired, pretty girl whose photographs show her wearing large-print, flowery summer dresses. In an illuminating video interview in 2011 she told her grandson Matteo, or Matthew,

"A tall, dark-haired, pretty girl . . .":
Posy and Dino in Brazil

that she met his grandfather when he was at MIT and was giving tennis lessons to Radcliffe undergraduates. One day a friend invited her to a Radcliffe party and Dino was there. He already knew of her but mispronounced her name. She was asked, "Is your name really Pussy?" She responded with a deadpan stare. Dino, she discovered, was very popular with the ladies and an impossible tease. "He used to tease me that to get him I had lied about my age," she told Matteo. "It turns out he was only a year younger than me." They had a small wedding in New Hampshire and then a honeymoon in the White Mountains before going to Brazil.

In 1941, when Posy discovered she was pregnant, they decided the moment had come to return to New England—Dino, by virtue of his marriage to Posy, had an American passport—and were en route to Andover, Massachusetts, by way of Trinidad, a British way station in the Caribbean. Dino was carrying a letter the British found suspicious. Since Dino was Italian-born, the British decided he was a Fascist and locked him up in a British prison. Posy continued on to the United States and then appealed to the American consul but met resistance because those authorities suspected Dino of being a Communist. Months of frustration followed. "It took a year to get him out," she said, as Matteo reported.

Ivrea had not, so far, been attacked but Milan and Turin were devastated by repeated Allied bombing raids. By the end of 1942 some 25,000 dwellings in Turin had been damaged or destroyed. In Milan the toll was even higher and half a million people left the city. Natalia's parents stayed on in Turin until their house was damaged as well. Giuseppe Levi refused to go to the basement during the raids. "My mother always implored him to do so, saying that if he would not, she should not go down there either. 'Silly nonsense,' he would say on the stairs. 'If the building goes, the cellar goes too. You can't pretend the cellar is safe. It is nonsense.'"

Just when people thought they were safe, Natalia Ginzburg continued, "suddenly bombs and mines exploded everywhere; houses collapsed and the streets were full of ruins, and soldiers and refugees. There wasn't a single person left who could pretend that nothing was happening—shut his eyes, stop his ears, or hide his head under the pillow—not one. That was what the war was like in Italy."

Somehow letters from America still arrived, and Camillo and Luisa received the news that Dino and Posy had had their first child, a boy. They named him Camillo. It was high summer 1941.

The proud grandfather wrote back at once. He wanted to make sure the new mother would be very careful, because "during summer, infants are very easily affected by belly diseases . . . it is necessary to give him very light, pure water." The water must first be boiled "and when it is cooled again, it must be aerated by energetically shaking the bottle." Only constant vigilance would do. Who knew what terrors might strike this precious baby unawares.

❡

Among the dramatis personae who enter the scene at this point is James Jesus Angleton of Illinois, who became one of the CIA's best experts in counterintelligence during and after World War II. As Angleton's biographer Michael Holzman and others have observed, Angleton liked to cultivate orchids in his spare time—not that he ever had much of that—and like the varieties he preferred, he was a hybrid. His father, James Hugh Angleton, was serving as a cavalry officer under General John Pershing during the campaign against the Mexican rebel Pancho Villa in 1916. In Nogales, Arizona, he met and married a seventeen-year-old, Carmen Mercedes Moreno, of mixed Mexican and Apache heritage, high-spirited and exceptionally beautiful. When Angleton senior left the military to join the National Cash Register company, an assignment that took him and his family to Milan a decade later, James, or Jim, as he is usually known, was sent to British public schools before going on to Yale. The British experience was, as Ben Macintyre, Kim Philby's biographer, wrote, formative. Angleton left with "courteous manners, a sense of fair play, an air of cultivated eccentricity, and a faint English accent that never left him." "His

James Hugh Angleton in uniform: a rugged American, 1944

major at Yale was English literature but he was more interested in poetry, founded a magazine called *Furioso,* and, in the course of soliciting contributors, met and became friendly with Ezra Pound, who was rabidly pro-Fascist and anti-Communist. After Angleton made a trip to Italy to visit Pound in the summer of 1939, just as war broke out, that led to an early investigation of the young man's politics by the FBI. According to his file, "Angleton advised bureau agents in 1943 that he was intimately acquainted with Pound . . . and subscribed to Pound's political theories."

It is said that Angleton's love of poetry uniquely fitted him for the task of ferreting out motives in search of the truth that would become his lifelong obsession. Poetry's hidden meanings and cryptic references—Angleton was particularly fond of an intricate poetic form called the sestina—require inspired deduction. So do the multiple messages any spy is bound to transmit and the oblique implications that may, or may not, be worth an interpreter's trouble. When he became chief of counterintelligence, biographer Ted Morgan wrote, "Angleton had to decide whether . . . defectors were genuine or plants; whether their information was true or false; whether, when one of them said there was a mole inside the CIA, he was telling the truth or trying to sow discord. The lesson . . . was that it's preferable to have unfounded suspicions than to be duped by a bogus source." The great irony of Angleton's career would be that the man who was given almost mythical status for his ability to ferret out the truth, and "turn" enemy agents to his own purposes, was also famously blind to the biggest "mole" in British history and the truth about a spy who was one of his closest friends.

All this was to come when Angleton, newly graduated from Yale, was drafted in 1943. People found him "a strange man, a genius." "With his jet black hair, expressive hands and piercing eyes, as well as emaciated good looks, he was enormously attractive to women," Robin Winks wrote in *Cloak & Gown*. Even so, there was a quality that did not invite amorous dalliance; he seemed "mired in loneliness. With his premature stoop, heavy glasses, long hands and fingers—the Goya look, one said . . . Angleton seemed to be hiding behind his work," Winks continued.

If ever Angleton had any doubts about his future direction, that was solved for him by the Army and the Office of Strategic Services, or OSS, the newly (1941) formed forerunner of the CIA. The morale of that group was, one insider wrote, "appallingly low." It had become a "convenient dumping ground for useless career officers," and overrun by "playboy bankers and stupid sons of wealthy and politically important families." In anticipation of D-Day the OSS was looking for recruits with a working knowledge of French. Knowing Italian did not hurt either and with his sophisticated cosmopolitan background Jim Angleton looked like a natural. He also had some powerful friends. James Murphy, head of X-2, a special branch dealing with counterintelligence, thought he would do fine. So did Norman Holmes Pearson, a former professor at Yale and an important voice in the OSS. More to the point, his own father was an advisor to X-2 and the senior Angleton was a committed anti-Communist.

Angleton was duly dispatched to an OSS training course during which long marches and obstacle course sought to instill in the newcomer the kind of sangfroid needed in a crisis. For instance, one exercise required the recruit to light a stick of dynamite, carry it unhurriedly to the middle of a field, and quietly walk away, displaying the utmost casualness while calculating to the millisecond how long it would take him (or her) to be safely out of harm's way before the explosion. History does not record how many potential pyromaniacs miscalculated. Angleton, who was primarily a strategist and planner rather than action hero, was not pleased. He had been instructed, he told Pound, about all the ways a man could kill or be killed. It was awfully boring. Such skills were considered essential because, as Thomas Powers, biographer of Richard Helms, the future director of the CIA, explained, it let the recruit know exactly what kind of business he was engaged in. That was "office routine at one end . . . treason, betrayal and violence at the other."

Again and again, the assessment of Angleton's personality and abilities would be the same. In *James Jesus Angleton, the CIA & the Craft of Counterintelligence,* Michael Holzman cites a fellow recruit who called him "extremely brilliant but a little strange. I met a lot of important Americans, from [William] Donovan on down, but Angleton was the personality who impressed me the most. . . . A very exceptional man. He had . . . a strange genius I would say—full of impossible ideas, colossal ideas." He came through training with predictable acclaim and was sent to London to serve under Norman Holmes Pearson, now head of X-2.

James Jesus Angleton in London during the Blitz, 1943

☙

"In this game London was the place to be in 1943–44," Winks wrote. Since the involvement of the OSS was essential in plans for D-Day, most of its western European operations were devised in the many empty and partly damaged Georgian houses that lined Grosvenor Square. The square itself was dominated by the American embassy, then headquarters for the Supreme Commander, General Dwight D. Eisenhower, and so many buildings had been requisitioned that it became known as "Eisenhower Platz."

At the end of 1943, when Bill Casey, a future director of the CIA, also arrived there, London had been subjected to such prolonged and concentrated bombing—300,000 homes destroyed and 20,000 people killed—that shattered windows, now boarded up, along with empty lots filled with rubble, attested to the price that had been paid. Even though the worst of the bombing was over, London still felt like "a city under siege," Douglas Waller, Casey's biographer, wrote. Casey's own command post was a five-story brick town house at 70 Grosvenor Square, one of the few still undamaged. Norman Pearson's group found equally elegant quarters that year in Ryder Street within a stone's throw of St. James's Palace and a short walk from Piccadilly.

From such rarefied haunts Angleton trudged back each night to a shabby neighborhood near Paddington Station and a bachelor flat. Both men would need to take defensive measures should they be on their way home when the air-raid sirens went off. This meant making one's way from doorway to doorway toward the nearest shelter, usually an Underground station, the further underground the better. Ever since the blackout everyone carried a flashlight, but this was of little use in a fog or one of the city's famous "pea soupers." To the sense of siege one had to add a nightmarish sense of dislocation; one entered a vast yellowish white blankness that might, or might not, disgorge headlights that turned out to be one's bus home.

Having arrived at the relative safety of one's digs one's attention might be caught by "the slow drumming sound" of an enemy plane overhead, then a barrage of gunfire. And then, as Elizabeth Bowen wrote in her evocative novel set in London during the Blitz, "down a shaft of anticipating silence the bomb swung whistling. With the shock of detonation . . . four walls of in here yawped in, then bellied out; bottles danced on glass." One's own fragile sense of safety had been shaken once more. Next morning there was always the acrid smell of singed dust and smoke, "the icelike tinkle of broken glass," the roped-off squares containing time bombs, "drifts of leaves in the empty deckchairs, birds afloat on the dazzlingly silent lakes," and the gnawing fear of what the next night might bring.

☙

In a matter of months the X-2 division had expanded from a skeleton staff to seventy-five people, and Angleton had risen dramatically from humble foot soldier to second lieutenant and chief of the Italian desk for the European Theater of Operations. He had also made the requisite liaison with the British Secret Service MI6 and its emissary Kim Philby, with whom he struck up a friendship over lunches washed down with an alcoholic fervor that matched Philby's own. Philby's specialty was how to "turn" enemy agents into double agents, reporting on what the enemy was doing, a skill about which he could happily talk for hours. One admirer reported, "He really knew what he was doing." Philby "may have felt he had a mentoring relationship with Angleton; Angleton may have shared that feeling."

Angleton needed to master "the central axiom of a new counter-intelligence philosophy," Tom Mangold, his biographer, wrote. "He learned that penetration of the enemy's intelligence services was crucial in order to nudge the enemy into . . . the famous 'wilderness of mirrors,' that mythical hell to which spycatchers are consigned by default, doomed to spend their working lives trapped inside the shimmering bars of glancing reflections." The rewards were few and demanded not only skill but patience: an ability to wait years for a successful result.

Angleton was not only learning how to deal with ambiguity and conflicting assumptions, but impressed by the British skill in uncovering Nazi spies and encouraging them to deceive their former masters with false information.

Once in Italy, Angleton and his co-workers would need to deal, Winks wrote, with "the incredibly complex problem of the Italian partisans, who were of a bewildering variety of political hues; how to keep those who were clearly Communists applying themselves to the common war effort, rather than diverting supplies and using information for postwar purposes; to help rebuild the Italian intelligence services, (which would be) hopelessly compromised by their cooperation with the Germans and almost as hopelessly fragmented in their views of the future Italian national interest . . . and bend all these secondary agendas to the principal goal" of defeating the Germans.

There were ever so many ways of turning problems into opportunities in disguise, if one knew how. Angleton began "studying the esoteric secrets of counterespionage" with single-minded purpose, "as if they contained the secret of the Trinity." He moved a cot bed into his office and became known for spending all night at his job. He was to be found in the small hours, "a single lamp burning, reading reports carefully turned face down on his desk . . . relaxing over poetry, smoking incessantly." He would become, as another mentor observed in 1985, not only unbelievably understanding of the nature of counterintelligence but "the finest counterespionage officer the United States has ever produced."

Angleton's father, James Hugh, was as assiduous as his son at tracking down a quarry, although his pursuit had more to do, at least at first, with financial advantage than intelligence gathering. At his Milanese offices Hugh Angleton was a superb host, tirelessly entertaining the circle of European friends he was acquiring in his role as vice president of the National Cash Register Corporation. The company was originally founded in Dayton, Ohio, to build mechanical cash registers. By 1911 it had sold a million machines, employed almost six thousand workers, and had bought out its early competitors to dominate 95 percent of the American market. As it expanded, NCR moved far beyond its relatively modest goal of making machines to track sales and foil thieves. During World War I, it manufactured fuses and aircraft instrumentation. During World War II, NCR would build secret communication systems, high-speed counters, and cryptanalytic equipment.

All that nonstop entertaining had a hidden importance for Hugh Angleton. Winks wrote, "From these visitors he received information on arms manufacturing, especially in Germany, and statistics on the duration of running time of various models of German engines, on fuel capacities and on flight distances between factories."

As he traveled to NCR factories in Germany, France, Poland, Romania, and Hungary, Angleton senior was creating what his son called "an internal spy trade" that turned out to be invaluable once war was declared. Hugh Angleton enlisted, joining the OSS under William

Vanderbilt, former Republican governor of Rhode Island, now executive director for special operations. Assigned to X-2, Angleton was sent back to Italy in 1943 and rumored to be working side by side with his son. Even if this was not quite the case there is no doubt that his father's extensive European contacts—he was also a former president of the U.S. Chamber of Commerce in Italy—were extremely useful to his son. Hugh Angleton became "very close" to William Phillips, U.S. ambassador to Rome. And he was an old friend of Thomas J. Watson, who had been a sales manager for NCR in the days before he founded International Business Machines. That was perhaps the most important contact of all.

In the time-honored saga of high sales and hard slogging, Thomas J. Watson, who ended up one of the richest men of his day, began life in the humble backwater town of Campbell in rural New York state. Perhaps some family gumption was inherited because Watson's parents had four girls before he, the fifth and last child, arrived in 1874, their only boy. After taking an accounting and business class, Watson started work at the age of seventeen, which would have been late for farm boys in his generation. He began as a bookkeeper for a local market for $6 a week. In an effort to improve that meager salary he started selling organs and pianos to farm families for the improved salary of $10 a week. He was a born salesman. Showing that business acumen that would serve him so well, Watson, who thought he could still do better, discovered that, had he worked on commission instead of a weekly wage, he could have made seven times as much. This first lesson in the hard truths of the business world would be turned to good account in the years to come.

By dint of determined self-promotion he finally wangled a job in the National Cash Register company, where he learned several other valuable lessons in sales and management survival. Before long he became the company's most successful salesman on the East Coast, earning $100 a week, a fabulous sum for the 1890s and unheard-of for a young man in his early twenties. The lessons learned, which also involved knocking out your competitor by fair means or foul, led to a rapidly advancing career at National Cash Register. In 1924, at the age of forty-eight, he founded his own company, International Business Machines.

The modest goal of selling cash registers so that small business owners could keep track of sales had developed into a much bigger and grander objective. Simply put, Watson was selling machines that made lists but on a scale and at a speed that had never been attempted before. The success of IBM is based on an ingenious method of punch cards that was invented by Herman Hollerith, a German-born young assistant to the U.S. Census Bureau. Hollerith had noticed that a train conductor could punch holes in a ticket in a special pattern to record details about a passenger's appearance, such as height, hair color, clothing, and so on. This was clearly an advance in the cause of preventing someone else from using the ticket. Hollerith, who was set to work on the census of 1890, came up with a brand-new idea. Why not use the same method to record gender, nationality, occupation, and so on?

"By virtue of easily adjustable spring mechanisms and brief electrical brush contacts sensing for holes, the cards could be 'read' as they raced through a mechanical feeder," Edwin Black wrote in *IBM and the Holocaust*. "Millions of cards could be sorted and resorted. . . . The machines could render the portrait of an entire generation—general or specific—or could pick out any group within that population. . . . Every punch card could become an informational storehouse limited only by the number of holes. It was nothing less than a 19th-century bar code for human beings." When the Census Bureau sponsored a contest for the best automated counting device, the Hollerith system won hands down.

The young inventor was quick to see the advantages of his new system. It could be adapted in all kinds of ways, from routine bills for a railroad to financial records for a huge insurance company. But his real stroke of genius was to realize that he should rent out his machines, not sell them. That way, millions of punch cards would have to go through the one system designed by him to accept them and his invention would be safe. It was so obvious, and so inspired, that Watson saw its logic and advantages immediately. Lessons learned were improved upon and enlarged once IBM had acquired the Hollerith system and began exploiting it with increasing success. The first complete school-time control system was followed by the first printing tabulator and then a printing press built to duplicate punched cards at lightning speeds. In 1928 IBM introduced an eighty-column punch card to double the information it contained. This remained an industry standard until the 1970s.

೮ః

The civil war in Spain and the arrival of fascism in Europe in the 1930s had, at first, given rise to a predictable reaction in the U.S., that of isolationism: a determination to stay out of European and Asian entanglements. Woodrow Wilson had made the case for American intervention in the first world war, 1914–1918, and the cost in American lives was high. In the 1930s nobody wanted that to happen again. But neither did big American corporations, who were happy to sell raw materials and equipment for both sides of any conflict. Standard Oil provided the fuel. Ford and GM built trucks and equipment, U.S. Steel and Alcoa sent critically needed metals, and American investors bought stock in I. G. Farben, the huge German chemical manufacturer.

Among the companies participating in this lucrative trade was Dehomag, IBM's 90 percent–owned subsidiary. Shortly after Hitler came to power in 1933, Dehomag was asked to assist the German government in its task of ethnic identification, with particular reference to Jews and Gypsies. Watson thought it was such a stunning business opportunity that he went to Germany in October of that year and increased IBM's investment in the German subsidiary by another million dollars, a hefty sum in those days. Thanks to the Hollerith machine the Nazis were able to increase their count of Germans with Jewish forebears from an estimated 400,000 to 600,000 to 2 million people.

Every time the Nazis invaded and took another country—Austria, Poland, Belgium, Denmark, Holland, France—the call would go out for the rental of IBM machines to identify Jews and Gypsies in the conquered countries. These also ended up as tracking, counting, and recording machines in the concentration camps. It was all very efficient and very rewarding for IBM's bottom line. Edwin Black writes that Nazi Germany became the second most important customer for IBM after the United States. When his groundbreaking book, *IBM and the Holocaust,* was published in 2001, it was heralded as a major exposé of a little-known but central aspect of the German annihilation of Jews in the Holocaust. *Newsweek* called it "An explosive new book. . . . Backed by exhaustive research, Black's case is simple and stunning."

In the 1930s, Watson was as starry-eyed about Hitler as others had been, notably Charles Lindbergh, William Randolph Hearst, and the Duke and Duchess of Windsor. According to Black, Watson met Hitler several times. On one occasion, when Watson was attending an opera house performance, the Führer, arriving late, entered the royal box, now emblazoned with a swastika. The audience of businessmen, including dozens from the U.S., leapt to their feet "amid roars, cheers and wild applause," and shouted "Sieg Heil!" while giving the Nazi arm salute. Watson's arm was halfway up before, as the story went, he caught himself just in time.

Watson also met Mussolini in those years and predicted a glorious future for Italy. "Evidence of his leadership can be seen on all sides. . . . Mussolini is a pioneer. . . . Italy is going to benefit greatly." The Italian dictator presented the American industrialist with an autographed copy of a photo of himself. Watson displayed it for many years on the grand piano in his living room.

The Brown Affair

As the war dragged on, Natalia Ginzburg, who was in Rome, ran into Adriano Olivetti one day on the street. He still looked a bit like a vagabond, as he did when they first met, with his "dragging, solitary tramp's walk." He was trying to overcome a natural reserve by consciously squaring his shoulders and looking straight ahead with his "steady, pure, cold gaze." There was nothing to distinguish him from the crowd; he looked like everyone else. That was not quite right, either. "At the same time he looked like a king, a king in exile."

At that particular moment she and Leone were living obscurely, they thought, in an apartment near the Piazza Bologna. Leone spent his days working in the offices of the clandestine newspaper *L'Italia Libera.* He always came back at the same hour but on this particular day he did not return. The following morning Adriano appeared at her door. Leone had been arrested, he said. The police would arrive any minute to arrest her as well. She must leave at once. He knew friends who would take them in.

She wrote, "I shall always remember . . . the great relief . . . I felt that morning on seeing before me the familiar figure I had known since I was a child, after those long hours of solitude and fear, during which I thought about my own people so far away in the North, and wondered if I would ever see them again. And I shall always remember his back bending to gather up our belongings scattered about the rooms—the children's shoes for instance—and his good, humble and compassionate movements. As we left his face had the weary look it had had when he came to our house to take Turati away—that fearful, happy look he had when he was taking someone to safety."

When Mussolini threw in Italy's lot with Hitler's Germany in 1940 (May 28) he might reasonably have expected the war to come to a rapid end. Countries were being mowed down under the relentless Nazi advance: Austria, Czechoslovakia, and Poland, then Denmark, Norway, Belgium, and the Netherlands. In May 1940 the Germans were poised to carve up France and marched down the Champs-Élysées in Paris a month later. Victory was inevitable and it would soon be time for the two dictators to start dividing up the spoils. When Italian troops marched north into the French Alps in June 1940, they met severe weather and developed frostbite but little opposition. That same summer Marshal Pétain signed a groveling armistice with his old enemy, Germany. Britain would be next.

Then the winds of war began to change direction. Italy attacked Greece but was ignominiously repulsed and had to be rescued by Hitler's armies. Other defeats followed in North Africa. In the spring of 1941 Italy lost Eritrea, Somalia, and Ethiopia, involving heavy losses of troops and equipment. Once again the Germans had to intervene. During the next two years Rommel's Afrika Korps and General Montgomery's Desert Rats waged a bitter, hard-fought war in the desert. In October 1942, Montgomery beat back Rommel at El Alamein, and by May 1943 the entire army of Italian and German troops had surrendered. Libya, under Italian control since 1912, was lost forever, along with thousands more troops and their equipment. But perhaps Mussolini's worst decision was to send almost a quarter of a million men to join Hitler's reckless invasion of Russia, badly equipped and inadequately prepared for the bitter winters. The results could have been predicted.

By 1942–43 the Fascists were losing public confidence at home, along with political power. Martin Clark wrote, "[A]s the war was being lost and the Fascist regime was collapsing, the political and diplomatic manoeuvres . . . became . . . intense. In the background was heard the distant rumble of guns in Libya and bombs on Milan, of strikes in Turin and food riots in Matera, of runs on banks and anti-Fascist

*Allen Dulles testifying
before the Senate, 1959*

congresses; in the foreground, in Rome itself, shadowy figures—some resolute, many fearful—held worried conversations and sent out oblique signals to friends and foe alike." One of them was Adriano Olivetti.

Italian captains of industry who collaborated did not have to deal with the kind of vindictive retaliation that followed the liberation of France in 1944. In the first place, the period of German occupation was relatively brief. But more to the point, Italians had been invaded and occupied for centuries, leading to that art of deceit, camouflage, dissimulation, and betrayal that Machiavelli so much admired. With what scrupulous courtesy and every sign of humble attentiveness did they listen to the commands of their conquerors, how anxiously and submissively would they await the verdict of their efforts. And how assiduously would they work behind the scenes to carry out acts of undreamed-of treachery and revenge, how expertly play double games their enemies could not even begin to imagine, let alone match.

This was the case in Ivrea. While responsive to the war effort, Adriano Olivetti and his managers covertly aided the partisans, fed anyone in the town who appeared at the workers cafeterias, and developed false papers that were miracles of cunning. All this could be expected of Adriano, whose role as protector was in a context that included not only friends, family, and workers, but the nation itself. On who knows what pretext he began to make repeated trips to Bern, Switzerland. He had made contact with the big man himself, Allen Dulles, head of the Office of Strategic Services. He was now an American agent, Number 660.

—

Allen Dulles, the master spy, who took up residence in Bern late in 1942, was already a veteran of the intricate games of espionage being carried out in this seemingly neutral country. Dulles began his career in espionage in World War I as a junior member of the U.S. legation where, as David Talbot, his biographer, writes in *The Devil's Chessboard*, he was already developing the veneer of charm, masking calculation, that would make him so successful in years to come. Affecting the manner of a dashing cavalry officer, the tips of his mustache honed to fine points, Dulles danced with aplomb, played expert games of tennis, and gave cocktail parties in an Art Nouveau fortress on the cliffs of Bern called the Bellevue Palace. He was a great hit with the ladies and popular for a casual ease that was so often mistaken for genuine friendliness.

A case in point is the anecdote Talbot relates when, during World War I, Dulles was working at the legation along with a pretty young Czech émigrée. Quite soon thereafter they were having an affair. One day he was advised by his British counterparts that the lady was passing secrets to the Czechs and Germans. This could not be allowed to continue. Dulles must have replied, "Leave it to me," or words to that effect. He arranged to take the young lady out to dinner "and afterwards he strolled with her along the cobblestone streets to an agreed-upon location, where he handed her over to British agents. She disappeared forever." Talbot also wrote, "Dulles was capable of great personal cruelty, to his intimates as well as his enemies. . . . He was [untroubled] by guilt or self-doubt. . . . He liked to tell people—and it was almost a boast—that he was one of the few men in Washington who could send people to their deaths."

From the OSS, Dulles would go on to become head of the Central Intelligence Agency (CIA) (1953–61), but that was yet to come when he entered Switzerland at the height of World War II. The only way to get there was to land in Portugal, travel through neutral Spain, then the south of France, nominally under Vichy control, and by train to Geneva. When Dulles left New York on the Pan American Clipper, November 2, 1942, he was in as much of a hurry as anyone can be who tries to be on time in the middle of a war. He knew that Operation Torch, the top secret Allied invasion of North Africa, was planned to begin on November 9. He was also aware that as soon as the invasion was known the Nazis would use that pretext to take over Vichy-controlled France and his route through France would be closed off at twenty-four hours notice. However, he had a week. That seemed long enough.

During World War II those who could afford it shuttled between New York and Lisbon on the small but luxurious flying boat that accommodated twenty-five people on the long and arduous journey across the Atlantic via the Azores, and also often stopped in Bermuda. Its seasoned crews navigated by the stars at sixteen thousand feet and needed to be experts in maritime conditions as well. For instance, planes could not take off on waves that were more than three feet high.

The plane landed safely in the Azores, but then bad weather set in. By the time Dulles had landed in Lisbon and then flown on to Barcelona, it was already November 8. He writes in *The Secret Surrender* that he was lunching with some Swiss friends at Port Bou on the Spanish frontier when a Swiss diplomatic courier came up to the table in great excitement. "Have you heard the news? The British and Americans are landing in North Africa."

At that point Dulles almost turned back. "If I was picked up by the Nazis in Vichy France, the best I could hope for would be internment for the duration of the war." One gathers that he rather enjoyed the prospect of melting into the crowd, running for cover, making contact with the Resistance, and being spirited across the frontier illegally, a "black" crossing as he termed it. He would go on.

All went well at first. When he arrived at Verrières on the French side of the border, he was amazed to find himself smothered in kisses by the natives, who somehow thought the war had ended. The train was bound for Annemasse on the Swiss border. This, as he knew, was going to be much trickier. He had been warned that a Gestapo agent in plain clothes would be stationed there. There was certainly someone answering that description who scrutinized the passengers' papers as, one by one, they filed through the gates at the station. He was the only person asked to step aside. The agent said nothing but a gendarme did. He was very sorry but the monsieur could not proceed. Dulles summoned up his best French "and made him the most impassioned

and, I believe, the most eloquent speech. . . . Evoking the shades of
Lafayette and Pershing, I impressed upon him the importance of let-
ting me pass. . . . I also let him glimpse the contents of my wallet."
The gendarme, it seemed, was not impressed.

Dulles spent a few anxious hours before the same gendarme re-
appeared. His train for Geneva was about to leave at noon. The man
whispered, "Allez passer. Vous voyez que notre collaboration n'est que
symbolique." (Get going. You understand, our collaboration with the
Germans is just a formality.) Dulles raced for the train with seconds
to spare. He learned later that the French had waited for the moment
when the Gestapo agent, whose routine never varied, left for lunch at
the nearest pub and his noon siesta. Dulles wrote, "Within a matter of
minutes I had crossed the French border into Switzerland legally. I was
one of the last Americans to do so until after the liberation of France."

Dulles took up residence in Bern in a handsome fourteenth-century
mansion in the Herrengasse, on a high ridge terraced with grapevines
and with the river Aare below. He wrote that the house was chosen
so as to "give cover" to visitors who might not want to be seen at his
door. This, however, sounds disingenuous, since Talbot points out that
his arrival was publicly announced in one of the papers and he himself
"wandered openly through the streets . . . in a rumpled raincoat and
a fedora cocked carelessly on the back of his head. He did not have a
bodyguard and he did not carry a gun. He met openly with informers
and double agents in cafés and on the city streets." Germans did their
best to double-guess his motives. They planted spies outside his house,
his cook turned out to be another German informant, and his Swiss
janitor stole carbons from the bottoms of his wastepaper baskets. All
he would volunteer was, "Too much secrecy can be self-defeating." Of
course he did not mean a word. It is clear that in various devious and
well-calculated ways, he was feeding deliberate misinformation to the
watchers while establishing unsuspected avenues of communication
behind the scene. His operations were certainly successful, particularly
with respect to new recruits and that sub-specialty of the spy world,
double agents, so useful because they already came equipped with

The Crown Prince and Princess of
Piedmont waving to the crowd, 1930s

all kinds of insider information. None of these was hard to find in this European center of political and financial intrigue, "a teeming espionage bazaar," as Talbot wrote. Meanwhile Dulles, with his candid smile and big-handed welcome, went on weaving his tangled web.

When Adriano Olivetti claimed to have friends in high places, for once he was not exaggerating. A fascinating and lengthy memorandum from OSS records, dated June 14, 1943, has been released under the U.S. Freedom of Information Act. It reveals that for months Olivetti had been actively plotting to remove Mussolini and planning his replacement, and that his co-conspirator was no less than the future Queen of Italy herself. Some time in the summer of 1942 he visited Princess Marie José Charlotte, wife of Umberto, Prince of Piedmont and heir to the throne, in their summer palace, a twelfth-century castle in Sarre that they had extensively refurbished overlooking the Val d'Aosta. The Princess was the third and final child of Prince Albert of Belgium and his wife, the former Duchess Elisabeth of Bavaria. She married Umberto in 1930.

Some accounts give the impression that Adriano and the Princess were alone in their plans to bring down the Fascist dictator, but this was hardly the case. The idea began with a young university professor, Carlo Antoni, a student of Benedetto Croce, philosopher, historian, and politician, and the group assembled around them was formidable. There were Count Nicolò Carandini, who became the first ambassador to Great Britain after the war; Manlio Brosio, a prominent lawyer and diplomat; Ivanoe Bonomi, a distinguished anti-Fascist; Luigi Einaudi, who became the second president of the Italian Republic in 1948; and a young diplomat from the Vatican, the future Pope Paul VI (1963).

Adriano's complicated plan involved members of the Italian intelligentsia in exile, specifically the friends in Giustizia e Libertà, now in Paris. A British secret service report, now released and in British government archives, continues that Olivetti thought it was important to establish "a nucleus outside of Italy to which the anti-Fascist elements in Italy can adhere. His idea some months ago was that there should be, in effect, two Italian governments, one outside which would assume an attitude of belligerency against the Axis, and a government inside Italy which would throw out Fascism, but assume an attitude of non-belligerency, or semi-neutrality, as he said the country was too exhausted to immediately be thrown into war against Germany." The two governments would fuse into one as soon as possible, but in the interim he proposed a national committee of refugees on the outside and anti-Fascists on the inside. He proposed to bring into the committee Luigi Salvatorelli, former editor of *La Stampa* and committed anti-Fascist, Dr. Ugo La Malfa, founder of L'Italia Libera and the Action Party, Carlo Levi, and Emilio Lussu, politician, writer, soldier, and another Action Party member.

Quite what the Princess thought of this rather Byzantine solution is not recorded. But it is clear that Adriano and the Princess liked each other and that he made numerous trips to the Sarre Royal Castle in the foothills of the Swiss Alps. She represented the newly evolving royals, with their easy informality, their spontaneity, and penchant for doing slightly outré things like driving their own cars. Her husband liked a formal court and the trappings of grandeur; she wrote books and went dashing off to concerts and festivals. Adriano's trips were interspersed with efforts to interest and include prominent Italians such as, for instance, Benedetto Croce, who was not very encouraging. He went to see the pope. He attempted to solicit the backing of the eighty-four-year-old Enrico Caviglia, a hero of World War I, but was equally unsuccessful. (The great man died in 1945.) Adriano had

Marshal Pietro Badoglio, center, in conference with Brig.
Gen. Maxwell D. Taylor, at left, and Mr. Montenari
of the Italian Foreign Ministry in October 1943

already visited Marshal Badoglio, who would become prime minister and usher in the country's first postwar government. The British government report continues, Olivetti "found him in excellent form and desirous of examining the whole situation and expressed great resentment against Mussolini. He asserted that he himself had no political ambitions. . . . He did not commit himself as to . . . whether he thought he could effectively take action." Badoglio was not about to move a finger until the Allies did first.

Everyone knew that King Victor Emanuel III ought to have been at the head of the conspiracy. But the tiny figure, barely five feet tall, had stood aside when Mussolini came to power and done nothing much since. "He was an intelligent man, but unimaginative, timorous and remote," Martin Clark wrote. "He was not likely to take any initiative if he could avoid it, let alone one that would risk his Crown.

He wrung his hands, and hesitated; perhaps something would turn up."

Umberto seemed to be as fearful of the consequences as his father was. The one daring member of the family—the Princess herself—had warned Adriano that the King would never act. Behind the scenes, once the extent of her political actions was known, she had been summarily told to stay out of politics. Shortly after that she and her children were moved to the small town of Sant'Anna di Valdieri southwest of Cuneo on the French Alpine border. Even that jumping-off point was not considered

The diminutive King Victor Emanuel III, 1935

safe after Mussolini was deposed and the Nazis moved in. She and her children fled across the Swiss border to safety.

The French border was never safe but the Swiss border offered safe haven, if one could get across it. The Swiss, in a permanent position of "armed neutrality," had closed their borders in 1942, which did not mean, for a time at least, that refugees did not keep coming. By the end of World War II, 180,000 of them had made the trip across and 67,000 of them were considered to be temporary visitors from border towns, expected to return to Italy and France once the war ended. Olivetti seems to have crossed using ten-day temporary visas, but even these would come to an end. This was a moment when any travel presented advanced discomfort, if not impossible odds. Passenger trains in wartime were few and far between. They were always late and sometimes did not appear at all, without explanation. Platforms were usually jammed. Once a train arrived, there was a stampede for seats, any seats. People would load suitcases into the corridors, then try to sit on them for hours at a time. There would be shunting stops and starts, blacked-out windows, foetid air, a lack of food and water, filthy facilities . . . Things had not changed much since World War I when that intrepid American journalist, Richard Harding Davis, left

Paris in 1916 en route for Greece. It was a trip that, in peacetime, might take six days. It took him fifteen. The traveler "learns, as he cannot learn from a map, how far-reaching are the ramifications of the war, in how many different ways it affects everyone."

Before arriving in Switzerland, Olivetti said he had made contact with all the Resistance groups: the Communist Party, the Partito d'Azione, the Partito Proletario per una Republica Socialista, and the Partito Christiano. Of these, Olivetti considered the Communists to be by far the best organized, and so did everyone else. Olivetti also said he had been delegated to represent all four groups (including the Communists and the Action Party, who had rather different agendas) "because all were glad to profit by his opportunity to travel and make contacts—not because he is a prominent opposition leader," the anonymous writer of the British report explained. "The main purpose is that these groups want to time their revolutionary action with . . . Allied plans, because a movement starting too soon would be crushed with catastrophic consequences, resulting in an eradication of a whole movement." When the right time to strike arrived, the Action Party could be ready in six days, the Communists in four. "On the other hand, they want it to start soon enough [so] that the liberation of Italy will not be seen as a gift from the outside, and purely a consequence of invasion—which would compromise the future of Italy, and of democracy there, making these people seem thereafter only agents of foreign powers."

Olivetti had a further reason for wanting to interest American authorities. He was forming an ambitious plan for the future of Italy which he would come to call Comunità, and which would absorb his energies for the rest of his life. The writer observed that one did not dare get him started on that subject or he would digress with such plodding detail that there was no stopping him. The trick was to get him to submit a written synopsis—of no more than six pages—on the pretext that the matter was too important to trust to a conversation.

Most of all, Adriano Olivetti wanted to meet Allen Dulles. It seems likely that the detailed memo written by an intelligence officer on June 14 was meant for Dulles and that the great man invited the newcomer for an interview the next day. Adriano was smuggled into a side door of the house in the Herrengasse in order to avoid being seen and photographed. No doubt Dulles received him in his best twinkly-eyed, professional back-slapping manner and convinced his Italian visitor, as he did so often with others, that he had a friend in high places. It also seems likely that Adriano, at some point, launched into the detailed plan for the future of Italy that had so alarmed his interviewer the day before. One does not know, but Valerio Ochetto, his biographer, states that he came away disappointed, without saying why.

One does not know what Dulles thought either but it may be inferred. What Olivetti probably did not know was that the U.S., in its zeal to bring about a new order for Italy, was backing a different man. He was Count Carlo Sforza, then in his seventies, a descendant of a distinguished Milanese family whose origins went back to the Renaissance. Sforza had fought on the side of the Allies during World War I and had opposed Mussolini at an early stage. When forced to flee from Italy Sforza emigrated first to Belgium, where he wrote a book about the inherent dangers of dictatorships in Europe, then to Britain and the United States. He became known as "the spiritual head of the Italian anti-Fascists." Sforza wanted a republic on the American model. Olivetti, as was clear, was pro-royalist. The OSS was about to whisk Sforza to Italy at the earliest possible moment. But in any event, what Dulles wanted was a double agent ready for action, not an ambitious, left-leaning Italian industrialist who wanted to impose on American policy his plan for a new Italy, ad nauseam. He might however have a limited usefulness. So for the time being he retained his informant's number of 660. Olivetti had already made a sporadic contact with British intelligence, i.e., its Special Operations Executive, or SOE, and been given the farcically anonymous name of "Mr Brown." During his visit with Dulles he was introduced to the OSS liaison with SOE, who in turn passed along the name of a certain Sig. Rossi, living in Milan. In reaching the OSS, Olivetti was told to go through Sig. Rossi.

How much use Olivetti made of the OSS contact is not clear. Both intelligence agencies were preparing for a July 9 deadline when the Allies would strike the coast of Sicily. At this point Olivetti appears to have turned toward the British side, where the response to his ideas was more enthusiastic. A figurehead royal family, with a democratically elected government, sounded just right to British ears. The British agent who had received a copy of Olivetti's OSS interview, and was forwarding it to London, added, "Brown strikes me as an energetic and gifted person who has shown great talent for organization in industry. If he is all right, as I think he is, he is our best bet by far." There seemed to be an impression, on the British side at least, that Olivetti could single-handedly engineer a coup and take Italy out of the war. The writer added, "We should back him all out."

As the last days of June rolled by, bringing "Mr Brown" and his ideas closer and closer to the secret deadline, coded telegrams flew back and forth between the SOE in Bern and London almost by the hour. These were frequently written by someone using the initial J., probably the head of the Italian section, Lt. Col. C. L. Roseberry, who was obviously making the decisions about whether or not to encourage the "Brown Group" as it came to be known. His reasoning, set out in some detail in an inter-office memo, makes interesting reading. He supposed that Adriano Olivetti, a Jew who was at one time "mixed up" in liberal politics, had to be shrewd and self-protecting to have become the head of a typewriter company that now had a near monopoly in Italy. It was perfectly possible that he was a Fascist supporter because his manufacturing company had been in a position to switch rapidly from typewriters to weaponry. "He might therefore be regarded as one of the many who have improved their fortunes under Fascism and now fear the possibility of ruin through an invasion," J. mused.

On the other hand, it was possible that Olivetti really was an anti-Fascist and had merely paid lip service to the regime while waiting for the right moment to act. He wrote, "I feel the motive behind all approaches from such individuals or groups is not a desire to assist us from inside in making our invasion a success, but rather to produce a collapse of the regime, so that surrender can be brought about without the ruinous results of an opposed invasion." However as a practical matter it did not make much difference what the motives were, if such an initiative could prevent a battle from ever beginning. So he also gave the idea his blessing.

Time, however, was of the essence. Mr. Brown, often referred to by his real name, agreed that a face-to-face meeting was the next step, which would coordinate the Allied invasion with simultaneous attacks by the combined Resistance groups. Precious time was taken up while Mr. Brown and the SOE representatives in Bern and London debated how an actual meeting might be arranged. Mr. Brown "was willing to come out of Italy to discuss plans, or to receive and house an emissary from us who can be accompanied by a wireless operator," J. wrote four days later, on June 21. It turned out that Mr. Brown had a country house just outside Ivrea, calm and isolated, beside the small lake of Viverone. He suggested that a seaplane pick him up and fly him to London. Alternatively, a plane could bring an emissary from London accompanied by the all-important technician to send and receive coded messages. Much time was wasted debating this idea. The lake, just two miles across, was deemed too small and larger lakes like the Lago di Varano or the Lago di Garda presented other problems. They decided against a night landing with a seaplane as too tricky. On the other hand, they might pick up Mr. Brown somewhere on the Italian coast near Genoa or even the Adriatic coast. Mr. Brown was all for having them send a submarine to get him but was politely informed that such conveyances were reserved for emissaries of higher status than his own.

One of the episodes of an enormously popular BBC television series, *Yes, Prime Minister,* lampoons the prime minister, played by Paul Eddington, his deviously manipulative Civil Service advisor, as played by Sir Nigel Hawthorne, and various cabinet members. In "A Victory for Democracy," the Foreign Office refuses to quash a repressive regime with a series of circular arguments, delivered with an arrogant lack of logic. This was a great joke as the F.O. had by then garnered a well-deserved reputation for its inability to act that had become something of a national byword. In the case of Italy it seemed that more than simple bureaucratic inertia was involved. Anthony Eden, who was working in close collaboration with Churchill, particularly detested Mussolini and by implication, Italy as well. He had been

*Anthony Eden, right, the immaculately attired British
foreign secretary, with Norman H. Davis, U.S.
ambassador-at-large, at an international conference, 1937*

stung by Mussolini's gibe, when the dictator was about to invade
Ethiopia and Eden was attempting to stop him, that the British
foreign secretary was "the best-dressed fool in Europe."

By early July, the SOE was beginning to cool on Adriano and his
vague assurances. "We do not feel," someone wrote, "that the facts as
we have them indicate that Brown has the substantial backing necessary for him to start such conversations, and the proposals are too
nebulous." In any event the matter would have to be passed on to the
Foreign Office where, it was implied, it would certainly be ignored.
No, military actions, which were taking place in six days, could not
be postponed for his benefit. Still, the agency was loath to let go of
him entirely. Maybe he could be persuaded to keep on working just
in case he came up with something useful.

Operation Husky, the invasion of Sicily, was planned for the night
of July 9–10 because the moon was favorable. It rose early, giving

airborne troops the light they needed for an easy landing and set at midnight, which was what sailors approaching a hostile shore preferred. The operation was the first Allied landing on the home soil of the Axis, and was prepared with extreme care. General Sir Bernard L. Montgomery would command the British Eighth Army and General George S. Patton Jr. commanded the American Seventh Army troops. Some 3,300 ships conveyed seven Allied divisions—two more than would make the initial Normandy landings the following year. Two weeks later the Allies had captured Palermo, and were bombing Rome. By July 25, Mussolini had been arrested and the Fascist government had fallen.

As the war historian Richard Lamb writes, Mussolini's arrest and the appointment of Badoglio as new head of state took Britain and the United States completely by surprise. In the months when Olivetti and others were making urgent overtures to U.S. and British agencies, Eden, for one, refused to listen and no efforts were made to take advantage of the situation. However, a great deal could have been done, as Lamb writes in War in Italy. When the crisis came, Germany had only one incompletely equipped division stationed in central Italy. Had the Allied and Italian naval land forces attacked on the mainland just then, "the German army in Sicily would have been trapped and unable to cross the straits of Messina." But "[a]s a result of the Allied failure to discuss plans for a coup with the anti-Fascists, the opportunity was . . . thrown away." Others have cited the British contempt, not just for Fascists, but Italians in general, considering them undisciplined and cowardly; Churchill thought they would have to earn their right to join "the company of civilized nations." Britain was not going to help them get there. Such an attitude had grim consequences. On September 8, Italy surrendered. Ten days later the Germans entered Rome and took over. The provisional Italian government lasted just forty-five days.

The tragic result of this perverse failure to act was a nine-month German occupation, from September 11, 1943, to June 4, 1944. It not only prolonged the Italian campaign, with all that this meant in terms of suffering and lives lost, but added to the threat faced by its Jewish population. As Susan Zuccotti wrote in *The Italians and the Holocaust,* Mussolini's racial laws did not come close to approximating the wholesale arrests and annihilation practiced by the Nazis once they took control. Even so, many who had benefited from the benign Italian habit of ignoring the law were now at risk from their German occupiers, including the Levi family, thanks to IBM and its endless, efficient lists.

When Natalia Ginzburg wrote her first novel, *La strada che va in città,* she was able to publish it in 1942, if under an assumed name. And when she and Leone were sent into a two-year exile in Pizzoli, a poor village in the Abruzzi, they, like Carlo Levi, were given remarkable latitude. In a 1944 memoir she celebrates its simple pleasures, such as a walk every evening, arm in arm in a snowy winter landscape, or watching her little dressmaker make *sagnoccole* for them, a dishtowel tied around her waist. Then there was Giro who ran the grocery store, looking like "an old owl, his round, impassive eyes fixed on the street."

After their release and return to Rome, Leone, a cofounder of the Partito d'Azione, picked up his anti-Fascist activities. He was again denounced, this time in November 1943 after the Germans had arrived. He was taken to the German section of Rome's notorious prison, Regina Coeli. That was in the Via Tasso, "an undistinguished thoroughfare of 19th- and early 20th-century apartment blocks and schools," David Laskin wrote, "with a crumbling arch at one end and the sanctuary of the Scala Sancta (the sacred stairs that Jesus trod) at the other." When he visited Rome in 2013, the via Tasso looked like a comfortable and convenient place to live, if undistinguished.

But during the German occupation, via Tasso 145 was the most feared building in Rome. "It was here in a charmless, smudged yellow apartment house that the SS and the Gestapo had their headquarters, their prison and their torture chambers. During the occupation the place was so dreaded that Romans never called it the Via Tasso. Instead they would say "laggiù" (down there), as in, "He was hauled off laggiù."

There, political prisoners and captured partisans were imprisoned in tiny cells with no beds, windows, or toilets. The building, now a museum, displays remnants of those who lived and died there: a sock embroidered with the words "courage my love," smuggled in by someone's wife or mother, a bloodied undershirt, or worse. There is also a portrait of Colonel Giuseppe Cordero di Montezemolo, celebrated for having undergone torture for fifty-eight days without speaking. Leone Ginzburg survived for two and a half months before dying under torture on February 5, 1944. He was thirty-four years old. His wife wrote, "When I confront the horror of his solitary death, of the

Wanda Soavi, Olivetti's secretary and companion, July 1943

anguished choices that preceded his death, I have to wonder if this really happened to us, we who bought oranges at Giro's and went walking in the snow. I had faith then in a simple, happy future, rich with fulfilled desires, with shared experiences and ventures. But that was the best time of my life and only now, now that it's gone forever, do I know it."

In the hours following Mussolini's arrest and Badoglio's appointment as prime minister it must have been small comfort to Adriano Olivetti to reflect that he had seen the dangers of inaction at close hand. In fact, when Italy was still, technically, allied with Germany, he was in Rome and living with his secretary. She was Wanda Soavi, who had worked for him in Ivrea, was also a passionate anti-Fascist, and had shared the same dangers as he shuttled back and forth to Bern in his efforts to rally the British and Americans to his cause. On the other hand, his SOE contacts were looking at him with new appreciation. A long inter-office memo in SOE files dated July 27 conceded that Brown had shown great prescience, not to say presence of mind, and had been extremely effective in organizing Resistance activities. He was turning his attention to Rome and had contributed (by one account) the munificent sum of £10 million to their activities. Since an invasion by the Allies on the mainland was pressing, why not use Olivetti as an emissary to Badoglio? Badoglio would know where the German resistance was likely to be light and help the Allies seize Rome. It made all kinds of sense to the practical minds at SOE. The Foreign Office in London turned it down.

Adriano, as usual, met Badoglio anyway, probably on July 28. But now he had some real concerns about how Badoglio would respond, with good reason as it turned out. Although Adriano was expert in negotiating the crosswinds of Italian politics that might render "on the out" people who were "on the in" the week before, even he must have been surprised when his words of caution were not received with the consideration he might have expected. There was a further complication having to do with the spy agencies themselves. The OSS and SOE had been in close collaboration for most of the war. But Dulles was not convinced that the SOE, which was in charge of dealing with SIM, the Italian spy agency, was being prudent enough. Dulles's suspicions proved to be well-founded. It turned out that SOE's prize agent, a man named Almerigotti whom they trusted implicitly, was not even named "Almerigotti"—he was a certain Dr. Klein from Trieste—and worked for SIM. Everything the British were doing went straight to SIM in a complicated game of double jeopardy and so did everyone else whom the British were naive enough to trust. That included their SIM contact, Sig. Russo. Sig. Russo thought his prime minister would not appreciate the doubts being expressed by Adriano Olivetti.

No sooner had Olivetti passed along his message on July 28 than he was arrested and imprisoned in Regina Coeli. Wanda and his driver were also arrested and sent to the same prison. He, and they, had been trapped. The irony is that Olivetti never knew that his own countrymen had betrayed him.

8

A Black Crossing

osterity has not handed down an account of the time spent in Regina Coeli by Adriano, Wanda, and their driver, Antonio Gaiani. One can make a safe guess that it was nasty and brutish but in this case, mercifully short. Adriano was released on September 29, 1943, two months to the day; Wanda and the chauffeur the week before. That he had been imprisoned was known to the War Office in Whitehall, London. In response to a Foreign Office query a certain F. H. Norrish wrote to a P. J. Dixon that Olivetti had been arrested during Badoglio's brief term. Norrish writes that a cable was sent ordering his release but it was not received in time. He continued, "efforts are being made to try and rescue him." That was another case of "too little, too late." The letter, dated October 18, was almost three weeks out of date. Colonel Roseberry, writing for the SOE, reiterated the confidence, not only of his organization but also of members of the Badoglio government, in Olivetti's gifts of organization and ability to unite groups in a common cause. By then, protecting Badoglio's delicate sensibilities from reproach had evidently become less pressing.

Fortunately Adriano was out of action while his country went from one crisis to another and the great powers battled it out, getting ever closer to Piedmont, the homeland he was desperately trying to defend. That was just as well because the headlong rush of events might well have defeated even his prodigious ability to cope. Mussolini had been interned in a remote winter sports hotel in the Gran Sasso d'Italia, a part of the Abruzzo Apennines. The hotel was seven thousand feet above sea level and only reachable via a funicular railway. The

Gabriele D'Annunzio, poet, dramatist,
and aviator hero of World War I, advising
Mussolini sometime during the 1930s

Germans decided to rescue him. They dropped over one hundred parachuters from troop-carrying gliders over the hotel. Eight of the gliders touched down on a rocky landing zone adjacent to the hotel and its commandos stormed the building. Mussolini was hustled into a small two-seater observation airplane on September 12 and was reunited with Hitler the next day.

On September 25 the newly energized Duce proclaimed a new "Repubblica Sociale Italiana," known as the Salò Republic, and took up headquarters a few kilometers away from Gabriele D'Annunzio's monumental estate, Il Vittoriale degli Italiani, beside the Lago di Garda, where he had died five years earlier. Since both men greatly admired each other, the choice of venue on Mussolini's part hardly seems an accident. To the same degree, Mussolini can hardly have misunderstood the reason why he had been restored to a kind of power by Hitler. Martin Clark observed, "Italy was no longer allied to Germany; she was occupied by her instead." Rome and the ministries were, for the time being, in German hands and only the northern half of Italy was in Mussolini's nominal control. Meantime the Allies were arriving from the south. Italy had acquired two governments. Olivetti had been betrayed by an agent working for Italian intelligence and now nobody knew whose loyalties were secretly engaged elsewhere. There were plenty of agents who might, or might not, be working for both sides at once, hedging their bets and feeding information to whichever government seemed to be winning the war that week.

As William Fowler has pointed out in *The Secret War in Italy,* the two years following the Italian surrender on September 8, 1943, were the most agonized of an increasingly bitter war that threatened to tear the country apart. "Some Italian soldiers who had been taken prisoner by the Germans were murdered. Brutal Fascist militia . . . —The Black Brigades—would hunt their fellow nationals who had become partisans, destroying the villages they believed had harbored them. As the tide of war moved progressively against Germany, the cruelty of both the Germans and Fascist troops increased. Special Air Service (SAS) troops attached to the partisans would see terrible vengeance."

At this point Olivetti managed to act with the caution and prudence he had not always displayed in the past. After their release he and Wanda remained in Rome for two months since he was confidently expecting the arrival of Allied troops. This did not happen fast enough to suit him (Rome was not liberated until June 1944) and so he and Wanda traveled on to Ivrea. There they discovered that the local carabinieri were looking for them so they moved on to Milan and stayed there until early in 1944. Again, they discovered that the police in Aosta were on their trail. It was time to leave again.

Wanda plays an important if unsung role in the galaxy of women who came and went in Olivetti's life. Unlike Paola she was not very pretty. The daughter of a Catholic mother and Jewish father, she joined the Olivetti company as a typist in 1934 when she was twenty-five years old and gradually rose to the position of secretary in Olivetti's new publishing company. Unlike Paola she was not fashionable either, and no casting director would consider her suitable as a hostess in a literary salon. On the other hand, as the selfless helper of a man on the run she was exactly right. She shared his postwar vision of a democratic Italy supported by a monarchy, and her vehement anti-Fascism matched his own. She not only typed his manuscripts but helped print and distribute anti-Fascist propaganda on a large scale. She was, at that moment, indispensable.

To the special circumstances of love in wartime one has to add the complicating factor in Italy that, for Adriano's generation at least, which had no recourse to divorce, "the principal purpose of married life is not the impossible satisfaction of adolescent love dreams, nor the achievement of romantic ecstasy, not the perfect fusion of two souls, but the foundation of a new family. . . . It is naturally desirable that husband and wife be happy in each other's company, but it is not indispensable." Adriano Olivetti's attitude in this regard may be seen as typically Italian. He made good use of quiet affairs, discreetly managed so as not to cause comment. However, once the relationship had run its natural course he was quite likely to drop the lady and go on to the next, something that only presented a problem when the next person was someone he also wanted to marry.

What perhaps distinguished him from his peers was that he felt obliged to assuage any hurt feelings by coming up with another lover for her. He was surprisingly good at this tricky maneuver, conducted with much delicacy and tact—he usually took the precaution of sounding out the next candidate first. To a surprising degree the discarded lady amiably accepted her new partner and all was well. This gentle approach hid a certain competitive streak. Adriano had been known to loudly admire the conquests of friends, brothers, and even his own son. But then, Camillo had been known to do the same although, in his particular case, he limited himself to compliments. It would seem that Wanda was more than just a wartime romance, brought about by propinquity and necessity. Even so, a painful ending could have been predicted.

In the first years of war Camillo, who was shielded from harassment by Mussolini, remained his usual feisty, demanding, and competent self. He also remained in charge. He took treatments for arthritis with his usual mixture of caution laced with skepticism, and had dismissive comments to make about mud bath cures, so-called, after one particular stay at a Swiss spa. He said the treatment exhausted him and made his brain feel like sludge. He affected a mustache and long, straggly beard and liked to wrap his ample outlines in a shawl. The result was to make him look a bit like a benevolent *zingaro* rather than the important man of affairs that he had become. As before, his agile mind was constantly on the move from one subject to the next. Having solved the problem of typewriters he was now toying with eternal questions of the cosmos, as presented by this interesting new approach, Unitarianism. He was all for setting up a church and was offering his friends wartime lodgings in the house on a quiet Milanese street that he had bought for that purpose. Until the war was over, here was this very handsome mansion with five main reception rooms, a kitchen, a quantity of bedrooms, and a large garden, perfectly empty. Would they like it? An answer is not recorded and perhaps the offer was declined, given the bombing campaign under way. (The house survived.)

Camillo continued to read widely, surprising everyone by taking up the novel. "Overall, I quite liked it" he said of *Madame Bovary,* his equivalent of high praise. He was less enthusiastic about Tolstoy's *War and Peace.* Perhaps understandably, he fretted about Luisa whenever she was taking one of the numerous "cures" in which they both indulged. He had just learned, he wrote, that diabetes could be cured by diet alone, well aware that she liked to sit around placidly doing needlework and had grown exceedingly stout. Whether she, as imperturbable as ever, took the hint is unlikely. Most of all he worried about keeping everyone warm and fed. He was particularly worried about "Princess" Mimmina, who seemed to have been vulnerable to coughs and flu. Her bedroom had to be kept warm no matter what. As for feeding everyone he was down to one cow, which was bad. On the other hand they had raised a fine pig, which, when slaughtered and dressed, had weighed 196 kilos, or 432 pounds. So although they had very little butter they now had lard. Not to mention meat.

In short, Camillo, now in his seventies, kept up the same demands on his energy that he had thirty years before, with the predictable result. There were times when he could not bring himself to move and just wanted to sleep. His son Adriano, who went through identical cycles of spurts and crashes, was sympathetic about what they both loosely called their "nervous breakdown." It was March 1943 and he had just recovered from the same thing, Adriano told his father. He arrived at the point when he could not keep his eyes open, which his doctor told him was the first sign of recovery. So he had slept and slept, done nothing else for two or three weeks and felt fine. Camillo should try the same. His father was soon back to normal. "I do not know if I told you," he wrote to Carlo Lizier, also an engineer, at the end of March, "about my pump for multiple lubrication?" Their latest model had given excellent results: it distributed oil at about 120 cubic centimeters a minute, which was sufficient for most uses. Camillo was now applying for a patent. Meanwhile he had decided to fast one day a week by staying in bed. It was very enjoyable and saved on food.

The events of September, the arrival of a German occupation and all that this implied in terms of increased surveillance and mass arrests, galvanized Camillo. Two days after Badoglio fled he met with a clandestine committee formed to work with the Resistance, in a state of high alarm, anger, and determination. Now was the time to take to the barricades, particularly if the German troops tried to destroy the factory. They were already sniffing around it, but workers and clerks were united in their opposition to the appalling combination, under Mussolini's new government, of hard-core Fascists and Nazis.

The sudden crisis was making Camillo ill and his wife and friends were very concerned. Something must be done. He and Luisa had to leave their small house in the modest neighborhood of Montenavale in Ivrea because the Germans would be after him sooner or later. Camillo protested, but the fact was that he faced problems even he could not solve. He was bundled up, probably into a horse and cart, given the shortage of gasoline, and taken on a long and winding trip up into the hills, the safe haven since time immemorial for the Piedmontese refugee. His destination was Biella to the northeast. As the crow flies the distance is only eleven miles, but the road winds and bends as it climbs, becoming almost three times longer, and to get there in such circumstances would have taken several hours. His rescuers stopped outside the town in a small village named Pollone where he was hidden with a peasant family. Perhaps by then the news had reached him that Adriano had been put in prison, which would hardly have contributed to his peace of mind. His world was collapsing around him. His physical condition continued to deteriorate. He died in a hospital in Biella early in December 1943, three months later. He was seventy-five.

In Ivrea they still talk about the spontaneous response the news of Camillo's death aroused in its citizens. Camillo's body was being buried in the Jewish cemetery of Biella. On the day of the funeral an exodus began. Workers and clerks, men and women, friends and neighbors, rounded up bicycles, carts, horses—or even walked—and began the grueling trip up the foothills of the Alps. There are no estimates of the crowd but tradition puts the mourners in the hundreds, if not a thousand. Dino Alessio Garino, an economist, professor, and author, noted in his biography *Camillo Olivetti e il Canavese* that the Germans could have wiped out the surrounding Jewish population in a single massacre of the men and women jammed into the tiny cemetery, had they been so inclined. It was cold and raining hard. Men removed their hats and stood silently, bare-headed, as the coffin was lowered into the ground, their faces wet with tears. Most of the family members were in hiding themselves. Luisa had been moved to the neighborhood of Vito Canavese in Turin, where she succumbed eight months later, in August 1944. But by then Adriano was in Switzerland.

Adriano had run out of options. By great good luck he and Wanda had been released from Regina Coeli a matter of days before the Germans took control of the prison. His trip to Ivrea coincided with his father's death and the likelihood is that he joined the mourners at the cemetery. Presumably he also visited the factory and his mother, which would have been the last time he saw her alive. Returning to Rome, he found that neo-Fascists were combing the city for people who had recently been released from prison so as to rearrest them. He and Wanda wasted no time in leaving. After arriving in Milan, more reports came in about police searches and another threat of arrest. They prepared for what Dulles called a black crossing.

Peter Ghiringhelli was a young British-born son of Italian parents deported from Leeds when Italy entered the war, and was living in a primitive village in the province of Varese near the Swiss border. Like other adolescents growing up on that particular frontier, he knew about the smuggling of Jewish families into Switzerland during the German occupation. In a fascinating memoir, he described the hurdles the refugees faced, in considerable detail. He wrote, "Guiding Jews to safety could be lucrative work, and this attracted professional smugglers, petty criminals and worse. The price for getting a Jew across the Swiss border ranged from 5,000 to 10,000 lire if the road was relatively easy, but it could be as high as 40,000 lire for a difficult and arduous mountain route. Avaricious guides would often threaten to abandon people in the mountains unless they paid more, and a few unscrupulous villains, paid by the Jews . . . would betray them at the border and claim the reward for capturing them." Agata Herskovitz, who nevertheless survived her internment in a Nazi death camp, described how she, her brother, and father were betrayed. In the summer of 1944 they were conducted safely over the mountains and had reached the border at Cremagna when their guides suddenly turned around and whistled. "At that instant floodlights came on and customs guards rushed out of a small barracks shouting 'Halt! You are all under arrest!' We were dumb-founded, incredulous."

Adriano knew that Gertrud, Massimo's German wife, was secretly helping bona fide rescue organizations in Milan that had led many across the border into Switzerland. One of them was probably under the aegis of the Committee for National Liberation, or CLN. This had been set up by Giuseppe Bacciagaluppi, a Milanese businessman with a villa on Lake Maggiore at Caldé, where the organization was based. "Bacciagaluppi was a powerful figure in the Resistance, but a very unassuming man," Ghiringhelli wrote. The CLN had worked out an ingenious system of guarding against betrayals: a note that was torn in half. One half would be given to the refugees, and a second to the guide. No money changed hands until the guide had produced the other half of the signed note to the CLN, and had his half verified. That method was working perfectly, and Adriano signed up.

Just north of Milan numerous pencil-shaped lakes, including Como and the Maggiore, along with many smaller ones, descend from the high Alps, following the steep Alpine cuts, and empty out on the border with Italy. In addition many Italian towns are literally skirting the border—including Como just north of Milan—leading to a borderline of dizzying irregularity as it zigzags in and out to accommodate towns, lakes, mountains, and historic artifacts. This makes it an ideal playground for creative feints and dodges as desperate men and women searched for new ways to cross. Thanks to the Swiss National Archives we can retrace the route Adriano and Wanda took when they left Milan on February 7, 1944, for Switzerland. The local trains from Milan to Como probably went fairly often—nowadays it is a commuter route—and the trip would not have taken much more than an hour. That leads to the Swiss border.

But instead of crossing their guide took them west, to the small town of Varese. From there they set out for a considerable hike going due north through the woods, taking refuge in a mountain hut near Bisuschio where, no doubt, they spent a cold, uncomfortable night. Very early next day they changed direction, going east and south. They passed through the pretty village of San Pietro, ending up near a tributary of Lake Lugano so early in the morning that the post office at nearby Stabio was not yet open. They waited patiently for that to happen, then declared themselves, signed a considerable amount of paperwork, and officially arrived. Their appearance—as head of a company, Olivetti clearly had the means to cover their expenses—was courteously accepted. They continued on by train to the frontier town of Bellinzona where, presumably, they checked into a plush hotel and had their first decent night's sleep for many weeks. They were safe at last.

As the daughter of a physician from Loschwitz and divorced wife of an aristocrat, Friedrich Ritter von Raffler, Gertrud Olivetti had been assured she was safe from police harassment. She was, after all, a German citizen by birth and had three German children—two sons, now doing military service in Germany, and a teenage daughter living with them in Ivrea. At her marriage to Massimo in 1934 her three daughters, Magda, Erica, and Eleonora, were aged eight, seven, and four respectively. This put the girls in a potentially delicate situation legally. According to the Nazis they were not Jewish although their father was. According to Mussolini's Italian Republic, they were.

This might have been a manageable situation but nothing in war is easy. The Italian police had discovered that Gertrud Olivetti was helping Jews escape to Switzerland from Yugoslavia, as well as Piedmont partisans. She was obliged to keep moving from place to place, showing admittedly false papers until the spring of 1944. At that point she was in Ivrea. She later told Swiss authorities that she had managed to give the German SS, or Schutzstaffel, which was hunting for Jews, sufficient explanations in Turin and Aosta but then they somehow learned she had been helping Jews to escape. She had no options left. In mid-April she, too, made her escape.

Gertrud's entourage included not only her three daughters by Massimo and her teenage daughter, Jenny, but a maid as well, since she liked to travel in style. Besides, she had four-year-old Eleonora to think of, who could not be expected to go on hilly hikes. Happily, she knew all the routes and chose one that must have been fairly foolproof. It took them by train to north Como, then by bus to Maslianco, a suburban community on the northwest corner of Lake Como. There she was met by a woman guide who found them beds for the night. Before dawn the following morning they walked to the Swiss border where another guide, this time a farmer, showed them how to get across. This took them to the edge of Chiasso, where they gave themselves up—the whole operation cost an extravagant 180,000 lire. Photographs taken by the authorities on arrival show Gertrud with a few hairs out of place, but otherwise smiling. Her daughters, identically dressed in broad-brimmed hats, dark double-breasted coats, and neat white blouses, look ready for a school outing. It took them four days.

One day in September 1943 Nicky Mariano, Berenson's constant companion, went down to the Piazza San Marco in Florence. "Military vehicles filled with German officers had just arrived in front of the

Italian military headquarters to take over control." She and Berenson had been offered refuge in Le Fontenelle, a Medici palace in the hamlet of Garregi owned by the Marchese Serlupi, ambassador to the Holy See, who was confident that the Germans would honor their tenuous diplomatic immunity. By then Count Ciano had been shot on orders of Mussolini. The news galvanized the émigrés on the hills of Fiesole as no other single event could have done. At last they knew how high the stakes were. As for Berenson, that inimitable host and conversationalist with the gift of the retort, he still wrote as if the war presented some kind of abstract intellectual challenge to be argued. He moved sideways with some reluctance. The Marchesa, trying to save him from himself, introduced him as "Monsieur le Baron" but was powerless to prevent him from going on daily walks and talking to anyone and everyone in his atrocious American accent.

Directly above La Piazzola, Paola's modest, half-hidden villa on a wooded hillside in Fiesole, the Villa Medici presented a tempting target for the German occupiers. As tales about the confiscation of furs, jewelry, and silver began to circulate, the Marchesa Iris Origo, Sybil Cutting's daughter, who had married an Italian, began packing up linen, blankets, and silver and finding hiding places in preparation for that eventuality. Owners of villas vacillated between their instinct to run for their lives and their desire to protect their belongings from wholesale theft, and worse. There were plenty of examples to cite, beginning with the Origos. La Foce, their country estate in southern Tuscany, was midway between Rome and Florence and in a direct line of the fighting. After the Allies triumphed Iris Origo wrote, "The house is still standing, with only one shell hole in the garden facade, and several in the roof. . . . In the garden . . . they have stripped the pots off the lemons and azaleas, leaving the plants to die. The ground is strewn with my private letters and photographs, mattresses and furniture stuffing. The inside of the house, however, is far worse. The Germans have stolen everything that took their fancy: blankets, clothes, shoes . . . and have deliberately destroyed much of sentimental or personal value . . . there are traces of a drunken repast; empty wine bottles and smashed glasses lie beside a number of my summer hats . . . together with boot-trees . . . overturned furniture and W.C. paper. . . . The lavatory is filled to the brim with filth, and decaying meat, lying on every table, adds to the foul smell."

In Naples it was the same story on a monumental scale. Norman Lewis, a justly famous observer, arrived there just after the Germans left as a member of the British Intelligence Corps in the autumn of 1944. He wrote, "The city . . . smells of charred wood with ruins everywhere, sometimes completely blocking the streets, bomb craters and abandoned trams." He continued, "It is astonishing to witness the struggles of this city so shattered, so starved, so deprived of all those things that justify a city's existence, to adapt itself to a collapse into conditions which must resemble life in the Dark Ages. People camp out like Bedouins in deserts of brick. There is little food, little water, no salt, no soap. A lot of Neapolitans have lost their possessions, including most of their clothing, in the bombings, and I have seen some strange combinations of garments about the streets, including a man in an old dinner jacket, knickerbockers and army boots." As Iris Origo asked a German officer when he was leaving for the front, he laughed and asked her just where she thought she was? She wrote that it was hard for civilians to get used to the idea that they were already in the middle of it.

Florence and its surroundings were particularly vulnerable for many reasons, the first being tactical. Because of Italy's topography—its Apennine chain of mountains tracing an undulating path from north to south down its center—it presented natural advantages to its defenders. The first defensive line was north of Naples and incorporated the monastery of Monte Cassino, a fight that took months with heavy casualties on both sides. The second, or Gothic Line, had been set up just above Pisa and Florence, running more or less directly above the Arno River from Massa Carrara on the Mediterranean coast to the seaside town of Pescara on the Adriatic. That, too, had been heavily fortified and so was Florence.

That battle promised to be another bitter fight, vastly complicated by the city's unique and irreplaceable store of paintings, sculptures, monuments, frescoes, and mosaics. The same could be said of cities all over Italy. In 1944 German Field Marshal Albert Kesselring remarked that he had not realized "what it was like to wage war in a museum." What could be moved already had been transported to the surrounding countryside, notably a bronze equestrian statue of Cosimo I de' Medici that was dismantled and removed from the Piazza della Signoria and reerected in one of the gardens on the Fiesole hills; many priceless pictures from the Uffizi and Pitti palace collections, moved for safekeeping to another hillside villa, had to be moved again when it was learned that the house was directly in the line of artillery fire.

In the summer of 1944, despite Hitler's declaration that Florence was "the jewel of Europe" and despite the mayor's desperate effort to save his city by calling it "open," i.e., unarmed, its German occupiers were not taking any chances. They blew up five bridges in Florence including the exquisite and irreplaceable bridge of Santa Trinita. They left only one, Hitler's favorite, the Ponte Vecchio, and this was made impassable by destroying everything on either side of it. German paratroop divisions took up defensive positions ringing the city. In Fiesole, to the north and east, residents who had not fled listened uneasily as the battle came closer and louder. On July 30, 1944, their electricity had been cut off and Berenson noted, "and nobody needs to be told what that means nowadays when so much depends on it. No more laid-on water, no mills going, no hygienic services, no more trams, no more radio. The telephone was cut off days ago. At last we were completely isolated. We can look over the battle that is raging, but without the interruption of the radio or even the newspaper we can make little sense of what we see."

Adriano had made sporadic visits and once he brought them all bicycles, which Lidia remembered. But now he was in Switzerland and the responsibility for the future of Roberto, Lidia, and Anna was on Paola's shoulders alone. Believing Roberto to be in most danger, she had enrolled him in a Catholic boys' school in nearby San Domenico; Lidia attended a day school run by nuns. It was, perhaps, her finest hour. The woman from whom too little had been expected, who spent her formative years reading books and dreaming about clothes and parties, would rise to the role required of her with ingenuity, tenacity,

*Roberto and Paola, presenting a united front in the garden at Fiesole,
about 1943. Anna, aged about five, can be partly seen at left.*

and fortitude. There is a photograph of Paola with Roberto in the
garden of their Fiesole house. It is undated but Anna can be partly
seen off to the side. She looks to be about five years old, which would
suggest that it was taken in the summer or autumn of 1943. The
impression is supported by Roberto's expression. He sits on a wall,
his arm around his standing mother's shoulders, leaning toward her
protectively, with a look of fear and defiance. Paola, smiling, eyes
half-closed, hands in her pockets, is perfectly coiffed, wearing beautifully tailored pants and a sweater, the image of casual nonchalance.
She would need every ounce of her superb self-assurance to bring
them through perhaps the most critical moment of their lives.

Giuseppe and Lidia Levi had moved in with them. Paola had chosen
to remain in place, waiting for the Allied arrival, and Lidia remembers
at least one occasion when Germans came to look over the house, and
Roberto and his grandfather hid in the attic. The small house was not
an early choice but i Tatti was, and two of its floors were taken over;

Mary Berenson, Bernard's wife, now bedridden, was moved to the attic, where she sweltered in the 90 degree heat. No one was living at La Pietra, the exquisite Fiesole villa sumptuously decorated by the parents of Harold Acton, the poet, novelist, and historian. They were safely in Switzerland, and Acton was on active duty in the Far East. The Germans also appropriated the Villa Le Balze, on the heights along with the Villa Medici, which had been designed by Pinsent and Scott in 1912.

As the British advanced, the net around their part of the Fiesole hills continued to tighten. Berenson's diary of the war in August 1944 is contained in *Rumor and Reflection,* one of his late diaries. It gives a day-by-day account of the bombarding and dynamiting as houses were blown up or ransacked, machine guns mounted, land ripped up to make trenches, and stray missiles threw splinters tinkling like glass among the potted *limonaias.* Berenson wrote, "From [our] dress circle, by moonlight yesterday, we enjoyed . . . a marvellous spectacle accompanied by . . . the growl, the rumble, the roll of cannon. . . . A distant mountain flamed up like Vesuvius," he wrote on July 31. "From beyond the hills came flashes of light, fan- or pyramid-shaped. This spectacle went on for hours."

The front rooms of Le Fontenelle faced the spectacular view over Florence, but the back of the house, which was carved into a rocky hillside, remained relatively safe. One day they were in these safe quarters when "an ominous rattle, clatter and swish made us start. . . . As I reached the drawing room I saw, gliding along the shut window blinds, what I would have taken for hissing snakes, if I had not recognized they were splinters of shell." The barrage of missiles went on for an hour. Afterward, "we were kept indoors more tightly than ever. Not only indoors but in the dark, for every opening was closed up with temporary contrivances. No reading, no writing."

No account has come down to us of the battle for Fiesole from the small family hiding in La Piazzola on the hillside below the Villa Medici. But one can imagine that the experience for Paola, her parents, and her children was essential in all respects that summer of 1944. It was only a matter of time before they, as Berenson feared, would be told to leave at once and loaded onto trains headed toward the concentration camps. A few days before Florence's bridges were blown up, in the middle of the fighting, Paola had a message. Happily, she had befriended a German official who gave her the advance warning that so often meant the difference between life and death. She must leave the next day before dawn. They were up and left at five a.m. Lidia, who was twelve and heartbreakingly pretty, with a mass of auburn hair, was just about to wash her hair when her mother stopped her: "We are leaving at once." She was wearing the only piece of jewelry she owned, an angel on a gold chain. Paola made her take it off for fear it might attract thieves, or worse.

As for Roberto and his grandfather, there was a large hospital in Fiesole treating several hundred of the wounded and refugees, and ambulances went down into the town every day. Roberto and Giuseppe donned white coats and became ambulance attendants for the occasion. The others made the perilous hike into Florence through the narrow back streets to reach asylum in the home of Paola's dressmaker. It took them two hours. It had been a terrifying experience but they were still alive and all of them were safe.

Before they left, the German troops planted delayed action mines, strategically hidden so as to explode when the owner started clearing away the damage. In the case of La Piazzola, sheets had been taken off the mattresses and everything was in a state of chaos, but at least nothing had been broken or stolen. They did notice that all the windows were left open except one. A long wire led from the single closed window to the back of the sofa, connected to a huge bag of explosives. It would have blown the villa and everyone in it to smithereens.

Not only villas received this kind of treatment. In southern Italy Norman Lewis was riding on his motorbike through a coastal area badly damaged by bombing when he noticed that the blue sky had become "a great opaque whiteness," rather like a pall of smoke, blotting out the landscape.

"On turning a bend, I came upon an apocalyptic scene. A number of buildings including a bank had been pulverized by a terrific explosion that had clearly just taken place. Bodies were scattered all over the street, but here and there among them stood the living as motionless as statues, and all coated in thick white dust. What engraved this scene on the mind and the imagination was that nothing moved, and that the silence was total. Dust drifted down from the sky like a most delicate snowfall.

"A woman stood like Lot's wife turned to salt beside a cart drawn by two mules. One mule lay apparently dead, the other stood quietly at its side, without so much as twitching an ear. Nearby two men lay in the positions of bodies overcome by the ash at Pompeii, and a third, who had probably been in their company, stood swaying very slightly, his eyes shut."

The scene was to be repeated all over Italy whenever the Germans left. Lewis, who was billeted in an old palazzo, faced the very real possibility that the same preparation for sudden death had been carried out below them. The building might blow up at any moment. Engineers were consulted but were not hopeful. Such a building was likely to be honeycombed with sewers, cellars, and the shafts of abandoned wells. "Even if there were mines, the odds were ten to one they'd never be found." The advice was to evacuate at once "and wait until the buildings stopped blowing up."

A Wilderness of Mirrors

On the other hand, remarkably few buildings in Ivrea needed to be replaced. That could be because, when Adriano Olivetti made contact with British and U.S. counterintelligence, he begged to have Ivrea spared from the attentions of America's "Flying Fortresses," the heavy bombers that were wreaking havoc on cities like Turin and Milan. He would naturally believe that his pleas had been honored. Or it could be that Ivrea, with its single employer, was not worth a detour. As it became clearer and clearer that the Nazis would be defeated, the ranks of partisans in the area doubled and tripled in numbers; 80,000 in March became 250,000 by the end of April. Grouped under the banner of the Committee of National Liberation for Upper Italy (CLNAI), the groups proclaimed a general insurrection throughout its area, Ivrea included, before the Allies arrived from the south. But dwindling numbers of Fascists and hard-core Nazis who had been chased into its Serra and Aosta valleys held on, and did not surrender until the news reached them that Hitler had committed suicide. That was on April 30, 1945. By then, the Allies had reached as far north as Biella. The war was over.

The Nazis had announced their intention of dynamiting the Olivetti plant as their parting protest. This never happened. Among the acts credited to the triumvirate of directors—Giuseppe Pero, longtime Olivetti employee, Giovanni Enriques, and Gino Martinoli—was an offer made to the senior German officer that he could not refuse. The occupiers took their bribes and left; the plant was saved. Given the fact that the Olivetti directors were actively helping partisans, the company and its employees were only slightly affected. Writing to his brother Massimo in 1947, Dino observed, "[O]nly three workers were taken to concentration camps in Germany. Only a very small percentage were shot by Germans as partisans." He did not give the figure.

Adriano, beaming from ear to ear, was given a hero's welcome and a town-wide celebration. He had a great deal to be grateful for to the men who had protected and defended the company and its workers, kept open house in the cafeterias, and saved many lives with fake identity papers. Meetings were held, speeches made and bonuses presented. Two of them—Pero, the company's indispensable financial officer, and Gino Martinoli, a first-rate engineer and brother of Natalia Ginzburg (who would soon return to his original surname of Levi)—were invited to stay. Enriques discreetly left. Adriano Olivetti was back and his family was in charge again. This reshuffling took place quite quickly in the summer of 1945. Two months later, in September, Adriano Olivetti left for Rome and Massimo took over as director-general. So just what did Adriano Olivetti think he was doing?

His year in Switzerland, and the sense of purpose he had found there, was formative. He had spent ten fruitful years developing the company's future. His ideas had not only turned a profit but helped train, house, and educate workers, offering them new opportunities and a new way of thinking about their future. Despite that remarkable achievement, he and his family were obliged to walk a fine line between their liberal humanitarian beliefs and the pretense of membership in the Fascist party they secretly opposed. During the war, the Olivettis, as Jews, had been persecuted by the Nazis and forced into exile. Adriano Olivetti must have reasoned that all his plans would be useless if political movements were going to wipe the company and Ivrea off the map. There was only one solution: he would have to start his own party.

Franco Ferrarotti, professor and author, who is considered the father of modern Italian sociology, was eking out a living in the post-war period working for Cesare Pavese, another anti-Fascist, whose writings reflect a uniquely romantic view of Communist beliefs;

besides several best-selling novels published in the immediate postwar years, he wrote some notable love lyrics. At that point, Ferrarotti was doing translations into English. He said, "Pavese was very interested in anthropology, mythology, ancient myth and had given me a book to translate by Theodore Reich on something to do with ancient ritual." After a stay in London in 1947, Ferrarotti returned to Italy and met Adriano at a party in Turin. He joked, "In those days I was always hungry and would go anywhere if there was something to eat. Anyway, I met this short man with an interesting, enormous head, a high forehead and these strange, clear, blue, childlike eyes. Innocent, and yet penetrating."

They began to talk and Adriano commented on the checked tweed jacket the young professor had bought in London. "He was very enthusiastic about the British, Churchill and Atlee in particular. He said that Churchill had won the war but the British realized he was no good for the postwar period." Ferrarotti, who considered himself an anarchist in those days, replied, "No, I am sorry sir, you are completely wrong. You do not understand anything," and they began to argue about whether life was better for the working man under Socialism, Adriano pro and Ferrarotti con. Then Ferrarotti said, "I have recently read that it is not enough to nationalize industry, you must socialize power." Adriano was completely won over. "That was when I realized one of his most important qualities was the respect and attention he gave to ideas."

Adriano had recognized a kindred spirit and wanted to hire him but events intervened and they did not meet again for several years. By then Adriano was making frequent trips to America. During one visit to New York he was interviewed by *The New York Times*. "Adriano Olivetti looks like an intellectual, mild Mr. Pickwick, with an exceptional high-domed forehead, curly blond hair and a somewhat rueful expression. Plump and comfortably sloppy looking, Signor Olivetti wears a bright blue suit, white shirt and white linen necktie as we lounge in modernistic leather chairs and sip pre-prandial cocktails." By then he was recognized as one of Italy's leading industrialists, whose company was known as much for its elegance of design as its function and reliability. He was in demand as a speaker. When he lectured at the University of Chicago, where Ferrarotti was now on the faculty, the reunion was cordial. Ferrarotti said, "he was trying to make his points but his liberal and progressive audience did not understand a word he said." The problem was his central thesis, which was an idea of democracy without political parties. "That idea was far beyond his audience, but what he anticipated turned out to be true, that political parties would become increasingly corrupt. His problem was that he was forty or fifty years in advance of his times."

Adriano was as dissatisfied with the parliamentary system as the Communist, because both centralized power. Ferrarotti continued that Olivetti's stratagem turned the idea upside down, i.e., that the power should be invested in the community. Unfortunately the name he would use for his movement, Comunità, was close to Communism, which may have made his distinctions and explanations less persuasive, for American ears, than they needed to be.

His "third way" began with small, self-governing communities, clustered around factories, laboratories, and universities. There would be "communitarian centers" hosting libraries and discussion rooms, as well as similarly designed social centers. These communities would be grouped into a consortium of municipalities. In this way no central authority would be able to impose its will on citizens living hundreds or thousands of miles away. Olivetti had anticipated the resistance that would follow the ever-growing size of the European Union. Such details were described at length in *L'ordine politico delle Comunità,* which spells out what he considered to be the right balance between centralized power and local autonomy, and his Movimento di Comunità would gain wide popularity in the areas of Piedmont bordering Ivrea.

Ferrarotti said, "Reform movements are either devoted to piecemeal policies, and they forget about the larger issues, or only to the dream and then they forget its practical applications. Adriano Olivetti was highly unusual in that he had a very practical approach to specific issues, but without ever forgetting a great dream. That is almost impossible to understand by today's politicians.

"He had an enormous craving, a spasmodic desire for power, the power to do things. People will never tell you this. He wanted the

*Adriano Olivetti in his
factory, postwar: "mischievous
and mysterious"*

power to do good. He was a marvellous man. And I share that because the secret of anarchists is that they love power! He was a unique figure. A practical, pragmatic idealist.

"One time he and I were giving speeches for Comunità together in Naples. We were staying at the Hotel Vesuvius one night when he came into my room. I was in bed and he was writing his speech for the next day. He knocks at my door. He wants my advice. 'Should I quote from the Bible?' I was tired. I told him I didn't know but if he felt strongly about it, he should. That seemed to satisfy him. I was almost asleep when there was another knock at the door. He had decided to take it out. 'I thought it was too strong.' Fine. He left. Twenty minutes later a new knock on the door. He wants to put it back in again. I say 'Please Adriano, anything you want, but please let me get some sleep!' He was like that.

"People found him mystifying. Some thought he was detached, aloof and dreamy. Some thought it was just a case of arrested development, a man who played at the idea of revolution because he could afford it."

Committed as he was, Adriano was discreet about his large ambition. "If you are very committed to an idea, a vision of the future, if your whole life is focused on that, you also think that any commitment that is exhibited will not be safe. It must be protected by discretion, shyness, and total devotion." Many famous politicians with good reputations respected his views but did not agree with him; his family even less. He did not seem to mind. "His idea was that, eventually, he could convince them all. I called it his prophetic complex. I told him, 'Why are you dealing with these people?' And Adriano would say, sotto voce, 'They have to understand.'

"I found him terribly attractive and complex at the same time. He could be mischievous and mysterious. I think he had some divinatory powers. This was a complex mind, half in love with complexity and using complications to cheat people, even to the point of somehow putting them on the wrong track as far as he was concerned. His love of complexity would make them think perhaps he was conspiring against them, a manipulator. On the contrary, he had complete respect for one's personal integrity. He would never fire anybody, for instance. Unlike Fiat, if you did not follow their political line. Never!" It was just that he reacted intuitively, especially when it came to sizing people up. "His secretary once asked him why he placed so much confidence in me personally. Adriano looked at him in some surprise. 'But look at the size of his hands!' he said. He thought the reason ought to be obvious to everybody." Ferrarotti paused. "I know he had a complete astrological chart on me." He had reached his own conclusion about Olivetti a long time ago and the importance of his humanist message in a consumer age. "This country did not deserve a person like Adriano Olivetti. He died too soon. What a tragedy!"

The hated Nazis had been defeated but the postwar situation in Italy continued to be chaotic. Badoglio had been succeeded by Ivanoe Bonomi, a prominent anti-Fascist who in turn was replaced by another anti-Fascist. He was Ferruccio Parri of the Action Party, the dominant voice in the Resistance groups in northern Italy, as a demonstration of the CLNAI's right to form the new government. Local partisan groups seized their opportunity to take terrible revenge on Fascist officials, factory managers, and fellow travelers. In the immediate aftermath of the armistice, twelve to fifteen thousand people were murdered in April–June 1945 alone. Apart from the appalling price paid in human terms, there was an economic price to be paid; the government was coping with the legacy of war, and it was higher than anyone expected. Martin Clark wrote, "Hundreds of thousands of ex-soldiers and former prisoners of war clamoured for jobs. Inflation reached record levels. Prices in 1945 were 24 times the 1938 levels, even after a freeze on gas, electricity and rents. . . . Over 3 million houses had been destroyed or badly damaged. . . . Industrial output in 1945 was about a quarter of the 1941 level."

In Naples, Norman Lewis found near famine conditions. He wrote, "Nothing, absolutely nothing that can be tackled by the human digestive system is wasted in Naples. The butchers' shops that have opened here and there, sell nothing we would consider acceptable as meat, but their displays of scraps of offal are set out with art, and handled with reverence: chickens' heads—from which the beak has been neatly trimmed—cost five lire; a little grey pile of chickens' intestines in a brightly polished saucer, five lire; a gizzard, three lire; calves' trotters, two lire apiece; a large piece of windpipe, seven lire. Little queues wait to be served these delicacies."

As for workers' conditions these were almost as stringent but bearable, since they and their families could at least survive, while daily demonstrations took place outside the gates of factories by the unemployed, frantic for jobs. The CLNAI were at work addressing old grievances. Piecework, i.e., paying workers by the job, was abolished. Rhythms of production were changed in response to workers' demands and a few national heads of companies as prominent as Fiat, in this case Gianni Agnelli and Vittorio Valletta, were declared persona non grata and subjected to commissions of inquiry. (They were later pardoned.)

Paul Ginsborg wrote, "The workers' desire to purge (*epurare*) undesirable elements, went beyond mere accusations of collaboration with the Fascists, to embrace charges that were of a purely class nature." Ginsborg cites the example of a foreman in a factory in Genoa who was considered guilty of "servile submission to capitalist directives tending to the hateful exploitation of the workers." Something of the sort took place in Ivrea in 1945. A group of leftists working at Olivetti held a secret meeting which resulted in the idea that the workers should take over the company. A certain Umberto Rossi, a Communist worker at Olivetti, was invited to do just that. The gentleman thought it over for a few days and then declined. He explained he would not know how to do that. Or words to that effect. There was no workers' revolution at Olivetti, or ever would be.

Ottorino Beltrami as a rising star in the Olivetti company. He is second from right. Roberto Olivetti is to Beltrami's right; another rapidly rising star. Gianluigi Gabetti is at far right, and Ugo Galassi, who had a distinguished career in the North American division, is second on the left, 1961.

Olivetti's work force was loyal but the speed with which many in and out of politics changed their allegiance was something to behold. Norman Lewis in Naples, then working for British intelligence, observed that his office was beginning to "receive a stream of visitors, all of them offering their services as informers. No question ever arises of payment. Our visitors are prepared to work for us out of pure and unalloyed devotion to the Allied cause." Their numbers came mainly from the professional classes, who left beautifully engraved calling cards, and titles such as Avvocato, Dottore, or Ingegnere. They had dignified manners and spoke in low, conspiratorial voices." Lewis wrote, "These are the often shabby and warped personalities on which we depend."

Ottorino Beltrami, the confidant who saw Adriano Olivetti off on a train that fateful day in February 1960, was of medium height, with blue eyes and a characteristically wide smile. A photo of him as he

walks toward a high-powered group of Olivetti executives in 1960
shows him with a smile of expectation, his only sign of nervousness
in his hands, fiddling with his cuffs. He was, like those who play for
high stakes, unassuming, the kind of person who blends into any
crowd and, as in the case of Kim Philby, the notorious Soviet spy, or
William Colby, director of the CIA during the Watergate hearings,
never lost his imperturbability. Most people liked Beltrami without
knowing him.

He could be most obliging. Gino Calogero, an engineer who
worked for him, called him "witty and smart" and admired his ability
to sidestep leading questions. Calogero recalled that Beltrami was once
asked why the Olivetti company had shut down but refused to say. He
"simply but cleverly said, 'Because I do not like to speak ill of other
people.'"

What people did notice, along with this discreet manner and comments that always ended in a half smile, was that Beltrami was "a
typical Tuscan." Perhaps they meant he was identifiable as soon as he
spoke because of his "giorgio toscana," a recognizable dialect in which
the letter "c," which in Italian is always pronounced as a "k" when
between two vowels, is pronounced with a soft "sh." There is also a
tendency to make greater use of the subjunctive—an indeterminate
form—than is customary in Italian. This may be inferred by the fact
that the author of *The Prince,* Niccolò Macchiavelli, was born in the
capital of Tuscany, Florence.

Beltrami was a graduate in engineering and naval sciences and
joined the Italian Royal Navy on the outbreak of war. In the beginning he was attached to several destroyers, rising rapidly through the
ranks. He then attended a submarine school in Pula and served on
submarines before assuming command of the *Acciaio,* a brand-new
submarine, in 1941 when he was just twenty-four years old. The vessel was launched in January of that year and delivered to the Regia
Marina on October 30. After delivery its crew underwent five months
of intensive training under the command of Lieutenant Commander
Beltrami while preparing for the submarine's first patrol north of
Libya. During this mission the new 700 HP diesel engine produced

by Fiat developed serious problems and the *Acciaio* was returned to the shipyard for refitting.

By June 1942, the submarine was back on patrol off the Algerian coast and the Balearic Islands. The *Acciaio* attacked a cruiser near Algiers but failed to hit its target. During its fifth mission, from January 1 to February 10, 1943, while on patrol between Cape Carbon and Cape Bougaroni off the Algerian coast, the *Acciaio* sighted and sank a British trawler. This was to be its only successful "kill."

References to Beltrami in biographical accounts frequently state that he lost a leg in a famous naval battle involving the British fleet and the Italian navy at Cagliari, the capital of Sardinia, in February 1943. Everyone in Italy knew about this attack, which inflicted major damage and loss of life. To say one had taken part and survived was tantamount, in Italy, to saying one had survived Pearl Harbor. In a loosely structured biography of his life, framed as a series of questions and answers, Beltrami is very specific about the day he lost his leg. He states that it was February 11, 1943. The problem is that, according to a highly detailed history of the submarine in *Submariners' World,* the *Acciaio* never saw action at Cagliari, and that sinking the British trawler was its only wartime success. On the day Beltrami claimed to have been wounded, February 11, 1943, the submarine was in dry dock.

The same account of the *Acciaio* records that the vessel resumed its patrols in mid-February for another tour of duty off the Algerian coast at Cape Bougaroni and also the Cap de Fer for two weeks in April. It was only at the end of this mission on April 16 that Lieutenant Commander Beltrami disembarked, leaving the command of the boat to Lieutenant Vittorio Pescatori. The latter was in command when the submarine was sighted and destroyed by a British submarine, the *Unruly,* in July 1943. All forty-six crew members lost their lives.

Beltrami also told his biographer in his book, *Sul Ponte di Comando,* that he received shrapnel wounds that turned septic—in the days before antibiotics—and his leg had to be amputated. One can forgive an old man a confusion about exact dates. But if he lost a leg it cannot have been during the battle of Cagliari, since, if we are to believe the historical account, the *Acciaio* was nowhere near it. There is the other matter of the shrapnel wound. How does the commander of a submarine vessel receive a shrapnel wound?

In some accounts but not in others, Beltrami is described as having been transferred to Italian Naval Intelligence once he had recovered from his injuries. If he received his wound in February, then it is conceivable that he was posted to Rome before Mussolini surrendered in early September 1943 and, by this account, crossed the line onto the Allied side, using a specially equipped bicycle, with forty-eight hours to spare.

On the other hand, if he was not wounded much before late April or May, the earliest he would have been fit for duty is late September or October, at which point he would have been reporting to Fascist intelligence and their Nazi overseers. Or, as already described, he would have been subjected to continually shifting loyalties, just like his many countrymen. The story is further clouded by the knowledge that in 1949 he spent at least a year working for the Marshall Plan. In her book, *The CIA and the Marshall Plan,* Sallie Pisani established the CIA's close involvement in the Plan and, by implication, its natural interest in an ex-Navy man who knew the Italian coastline. Most people who have tried to piece together a reliable account of Beltrami's life agree that there is very little information about his activities in the years before 1950, when he is believed to have joined Olivetti. It is possible that this accomplished naval officer with the disarming smile had worked as an intelligence operator for Nazi as well as Fascist overseers before his work for the Marshall Plan. There, as is clear from Pisani, he would not only have provided valuable information about past contacts but worked closely with the CIA, perhaps even the Mafia, although this is not clear. As Norman Holmes Pearson, that sophisticated advisor and observer, once said, "It is well in all cases to go on the old . . . axiom: 'Once an agent, always an agent—for someone.'"

As James Jesus Angleton also knew, it was almost embarrassingly easy to "turn" former enemy agents. In this case he was further helped by his father, now a lieutenant colonel in the OSS, who had served as its representative with Pietro Badoglio, and also to the numerous intelligence services. Writing in the *Executive Intelligence Review,* Allen Douglas observed that the "outspokenly pro-Hitler, pro-Mussolini senior Angleton also headed the U.S. Chamber of Commerce in Italy, and had extensive contacts with Mussolini's intelligence services. Some accounts report that he was a business partner of Allen Dulles."

As for his son, two years after arriving as chief of the X-2 unit in Rome and now chief of all OSS counterintelligence in Italy, he found himself in charge of Italian police and all of the military intelligence and secret services, since they all reported to him. This gave an extraordinarily far-reaching power and influence to someone who was still in his twenties. Douglas continues, "Essential to Angleton's activities, to the establishment of the first stay-behind units in Italy, and to the organization of the Vatican-linked 'rat lines' that smuggled Fascists out of Europe at war's end, was the Sovereign Military Order of Malta (SMOM)." This curious and little-known group whose members were drawn from Italy's so-called Black Nobility, also found new opportunities in the formidable empire that Angleton, one of the few to make the transition from the defunct OSS to the new CIA, was developing.

Angleton's willingness to work with Fascists was something that his immediate predecessors had declined to do on the perfectly reasonable assumption that it would be folly to trust a former enemy who might also be reporting for the other side, in this case Russia. In fact, this had already happened. Under Angleton's predecessor, Robinson O. Bellin, the OSS had recruited a number of Italian naval saboteurs, one of whom was discovered to be a possible German agent, and the operation was shut down. Angleton was willing to take the risk. In pursuit of information by fair means or foul, he was even willing to go as far as rescuing a certain Junio Valerio Borghese.

This curious and sinister figure belonged to a distinguished family in the "Black Nobility." As with the Pallavicini, the Colonna, and the Orsini, the Borgheses could claim numerous popes and cardinals in their family tree. Borghese, the so-called Black Prince, was a naval commander, a fanatical Fascist, and, as Daniele Ganser observed in *NATO's Secret Armies,* among the most notorious Fascists ever recruited by the United States. "As commander of a murderous anti-partisan campaign under Mussolini during the Salò Republic, Borghese . . . with . . . his special force of 4,000 men . . . specialized in tracking down and killing hundreds of Italian Communists," Ganser wrote.

As the war ended, Borghese was captured by partisans and about to be hung when word reached Angleton. Something had to be done, and Angleton volunteered to personally rescue him. One of his proudest boasts became that, single-handedly, he had saved the Black Prince, disguised in an American Army uniform (presumably tailored to fit), and conducted him to Rome. There Borghese was tried by an Italian court and convicted. He was sentenced but then, as so often happened in Italy, somehow released. Angleton received the U.S. Army Legion of Merit for this and other derring-do in 1946. Why, after all the blood that had been shed to defeat Mussolini, would a prominent officer of U.S. intelligence deliberately save the life of one of Fascism's most violent and fanatical practitioners?

Behind all the political maneuvers, the sordid deals that made friends of foes and that spawned a thriving backroom market in blackmail, double dealings, and betrayals, there was an inescapable reality. Another kind of war was shaping up, one that never actually became hot, but that lasted for decades, until the famous moment when President Ronald Reagan, in West Berlin in 1987, asked Mikhail Gorbachev to "Tear down this wall!" Barely two years after World War II ended, that malign phenomenon called the Cold War was taking shape and an "Iron Curtain" was descending across Europe. Former prime minister Winston Churchill was referring to the fact that, in order to protect Russia's border, Stalin was establishing Communist regimes in Poland, Hungary, Romania, and Bulgaria, with troops if necessary, as it would do in Czechoslovakia in 1948. The Russians were firmly ensconced in East Germany and seemed prepared to drive out the British and Americans from their respective sectors in West Berlin. After North

Korea invaded the south in 1950, the U.S. became convinced that the Soviet Union had embarked upon a worldwide strategy of aggression.

As Churchill described it, the demarcation line ran from Stettin on the Baltic Sea to Trieste on the Adriatic. In other words, Italy was on the front lines. Nuclear war loomed on the horizon, and America's brief monopoly ended in 1949 when Russia exploded its own atomic bomb, making the stakes even higher. Things could not have been worse for Italy, which veered between a policy of low-key neutralism and an effort to maintain relations with Moscow, hedging its bets. This could hardly have inspired its new Western allies with confidence. Its role in World War II had been, at best, ambiguous. Its politicians seemed impervious to the distrust, even contempt, with which it was held by its conquerors. As far as the victors were concerned, Italy had been defeated, was militarily helpless, and would need to make considerable amends in order to be trusted once again.

There was the immediate problem of a strong Communist presence in Italy as represented by its party strength and its many trade unions. Some crucial elections were coming up in 1948. In *Legacy of Ashes,* Tim Weiner's magisterial history of the CIA, he writes, "The CIA told the White House that Italy could become a totalitarian police state. If the communists won at the ballot box, they would seize 'the most ancient seat of Western Culture.' " That is to say, the Holy See. George Kennan, the CIA's Russian expert, thought that "a shooting war would be better than letting the communists take power legally—but covert action modeled on communist techniques of subversion was the next best choice."

If war with Russia was inevitable, the members of the newly formed North Atlantic Treaty Organization knew they were at a disadvantage because they were outmatched by the vast conventional forces of the Soviet Union and its allies. What was urgently needed, Weiner writes, was "a network of stay-behind agents—foreigners who would fight the Soviets on the opening days of World War III. The goal was to slow the advance of hundreds of thousands of the Red Army's troops in Western Europe." Hidden caches of arms, ammunition, and explosives would be stockpiled all over Europe, bought and paid for by NATO. One of the first such armies to be established was in Italy, called Gladio ("The Sword"). What was not revealed was that Gladio was recruiting ex-Nazis, Mafiosi, and ex-Fascists. They might be war criminals but they were also fanatical anti-Communists and therefore the new bedfellows. Borghese's name was on the top of Angleton's list. Borghese was always addressed as "Il Comandante." So, by a curious coincidence, was Ottorino Beltrami.

The secret armies established all over Europe "undoubtedly began as an effort to create forces that would remain quiescent until war brought them into play," John Prados wrote. "Instead, in country after country, we find the same groups . . . originally activated for the wartime function beginning to exercise their strength in peacetime political processes. Even worse, police and security services in a number of cases chose to protect the perpetrators of crimes to preserve their Cold War capabilities." As a result such networks continued to function with impunity, "long after their activities became not merely counterproductive, but dangerous." They were armed, ruthless, and no one dared bring them to account. The existence of Gladio was not known in Italy until it was revealed by Prime Minister Giulio Andreotti in 1990. More than twenty-five years later, many Italians still do not seem to know anything about Gladio. As for Adriano, it is impossible to know how much he knew about the developing Cold War or whether he thought there was anything to be concerned about. He had, after all, acted as a spy for both the British and the Americans. One of his closest colleagues, Beltrami, who was presumably working for him, would surely know what was happening behind the scenes. It could be that this lulled him into a false sense of security and a naive belief that he was now secure from whatever battles lay ahead. After all, the Marshall Plan had awarded Fiat with huge contracts even though it had been a pivotal factor in supplying Mussolini's war. How could an office supply company possibly have anything to fear?

ᚉᚔ

Adriano Olivetti was back in business, ready for action, and in the only place he wanted to be just then, which was Rome. The company offices were in the Piazza Barberini, near the Spanish Steps, and the first order of business was getting his magnum opus, *L'ordine politico delle comunità,* into print. There were lots of other works as well. So naturally he established his own printing house. He also started publishing a new magazine as a vehicle for his ideas. Ignazio Silone, novelist, author of *Fontamara* and *Bread and Wine,* and a pronounced anti-Fascist, was the latest rage and so it was gratifying to have him write the opening essay. His imprimatur would no doubt reach the intelligentsia in short order. At the same time, the government was in such political flux that it was, for Adriano, even more of a gamble than usual. As Allen Douglas wrote, postwar politics in Italy was "a wilderness of mirrors, with its rapid changes of government, multiple coup attempts and spectacular outbreaks of terrorism." In the 1946 elections the Christian Democrats, headed by Alcide De Gasperi, a professor, journalist, and the party's founder, won 35.2 percent of the vote and 207 seats out of a total of 574. The Communists, led by Palmiro Togliatti, who had returned from Moscow, won 104 seats, and their Socialist colleagues, 115. Between them, they constituted a plurality at 39.7 percent, but they had each fielded their own candidates and so the Christian Democrats won.

After a referendum the monarchy was abolished and King Umberto II stepped down. A new constitution was in place, which should have clarified and simplified matters. But no sooner was it ratified than new rifts developed to further fragment the parties. Four Communists who had been awarded government posts were forced out. The battle of 1948 was looming. Under the new banner of the Popular Front, the Communists and Socialists intended to unite in support of Togliatti, making victory all but certain.

The 1948 elections in Italy are considered one of the most flagrant interventions in foreign affairs ever conducted by the U.S. government. Catholic families with Italian backgrounds were urged from the pulpit to write letters to their Italian relatives, begging them not to support the godless Communists. Hollywood stars waved and smiled, urging the same message. The U.S. administration increased its "interim aid" to Italy to the tune of $176 million. As each ship, containing food, medicine, and supplies, arrived at a different Italian port, U.S. ambassador James Dunn was at the dock for a new photo opportunity.

A month later the Christian Democrats soared to an increased majority of 305 seats out of 574, or 48 percent of the popular vote. The Popular Front dropped to 31 percent. That particular crisis was over for the moment at least. And no one who mattered could fail to understand who Italy's new masters were.

As all this was happening, the Pentagon received an urgent message from the chief of American forces in Berlin, Gen. Lucius D. Clay. The general's intuition told him that the Soviets were about to attack. Tim Weiner writes that President Harry Truman spoke at a joint session of Congress the next day, warning them "that the Soviet Union and its agents threatened a cataclysm." He asked for, and got, the Marshall Plan. During the next five years Congress would appropriate some $13.7 billion to help repair the damage done by war and build a "political barricade" against the Soviets. Not only would the Marshall Plan act as a conduit to provide the CIA with untraceable amounts of cash. The new Plan would also act as a front for covert programs to recruit foreign agents, place propagandistic articles in newspapers and magazines, and conduct surveillance. The State Department wanted to spread rumors, use bribery, and set up non-Communist fronts. As for the Pentagon and the forceful secretary of defense, James V. Forrestal, they wanted "guerrilla movements . . . underground armies . . . sabotage and assassination."

After some months in Rome it might have become evident to Adriano Olivetti that this was not the moment for a man with Socialist views to launch a movement decentralizing power, however lofty and well-intentioned. There were practical reasons as well. Parties spring up like mushrooms before any election, but the postwar group was a special case, as Norman Lewis observed in the early summer of 1944. "There are now some sixty officially recognized political parties having memberships ranging from a hundred or so to nearly two million. Many of these offer bizarre recipes for national salvation, including a small band of fanatics in the Salerno area which claims to have discovered the solution of the problem of perpetual motion, and to be ready to exploit this in the national interest. In addition . . . there are clandestine Neo-Fascists and Separatists. . . . The Separatists' latest plan for Italy's regeneration includes the immediate demolition of all factories, the abolition of the motorcar, and the renaming of the months of the calendar after the Roman gods. This is the season . . . when insanity has become almost respectable."

Returning to Ivrea as he did in winter, following the 1946 elections, presented Adriano Olivetti with a new set of problems. If he had expected his brother Massimo to yield his role as president with a graceful smile, he discovered otherwise in short order. It seems likely that he refused point-blank. In this, Massimo was supported by Carlo Lizier, his late sister Laura's husband and Mimmina's father, who administered and controlled his daughter's one-sixth share of the Olivetti family fortunes. In such cases one finds wounded feelings on both sides. Adriano had been selected by his father to run the company and had not only done so successfully, but was ready to take it to new heights of popularity and renown. Massimo, as the younger brother—barely by a year—had been just as valuable, inventing new machines to retain the company's competitive advantage and playing a vital role in ushering through ideas that Adriano, who did not have the same gifts, probably hardly understood. An equal partnership might have been the solution. As it was, Carlo Lizier loudly supported Massimo. Letters flew back and forth, revealing hurt feelings on Massimo's side and a pugnacious willingness to take on all comers from his wife. Nobody liked to mention it but she was, after all, a German, despite her sterling work helping refugees. Perhaps the family felt it was appropriate for relatives by marriage to stand back and let the children of Camillo decide. Their verdict was a foregone conclusion.

The bright-minded, quick-witted Dino enters the picture at this point. The youngest member, who had spent the war years in Connecticut, where he and Posy were bringing up their young family, had the detachment distance can bring and clearly saw, he thought, the importance of having Adriano continue in control. It seemed necessary to make the trip to Ivrea (a nine-day crossing by ocean liner in those days), bringing Posy and the children with him, and stay at the Convent, which had been deeded to him in Camillo's will. But that meant that Gertrud, Massimo, their assorted children, and her mother would have to move out. While assuring Dino and Posy they were ready to leave, somehow Massimo and Gertrud stayed on, which naturally led others to wonder why. Adriano offered Massimo the position as head of the Spanish division of Olivetti. Massimo said that Gertrud would not move to Spain. Massimo said he would leave but he wanted a very big financial settlement for his many patents. Adriano hesitated. Lawsuits were filed. Torn between Adriano's refusal to yield and Gertrud's fury, Massimo became ill. Then he was bedridden. Gertrud wrote more letters, seeking to place the blame for his illness on Adriano. Some kind of settlement was being worked out when, on February 20, 1947, Massimo's heart failed. He was just forty-nine.

David Olivetti recalled that his mother had described to him what happened on the way to the funeral—he, being just six years old, did not remember. Posy said that he and his brother Alfred were walking to the funeral with Adriano, one on each hand, and passed by the Convent, where Gertrud was still ensconced. Suddenly an upper window flew open and Gertrud leaned out, dressed in a negligée. She was clearly not going to attend her own husband's funeral. "Goodbye!" she shouted, and that was that.

Enigma Variations

It is said that Camillo Olivetti, with his habit of roaming the halls of his production lines at all hours, came upon one of his workmen in lively conversation with someone who, from his hat and coat, clearly signaled a visitor who had not been announced. He went up to the stranger. "Who are you?" he demanded. The man turned with a smile. "I'm a psychologist," he explained. He was hustled out of the building.

The story is perhaps apocryphal, but it neatly illustrates one of the ways in which father and son differed. Camillo was on the alert for strangers who might, or might not, have sinister motives. Adriano, on the contrary, welcomed them, the stranger the better. Who knew what a fresh eye and mind might contribute? He would often turn loose newcomers from unlikely fields with the task of recording their impressions and recommendations. What made him interesting, Natalia Ginzburg thought, was his open-mindedness. Adriano Olivetti, she wrote, was "ready to throw out whatever he'd chosen to work on the day before. [He] was always anxious and restless in his quest to discover the new, a quest he deemed more important than everything else, and would stop at nothing to achieve it: not at the notion that he owed his fortune to his previous inventions, nor at the confusion and protests of those around him who'd become wedded to the old conventions and couldn't understand why they should be tossed aside."

One of the early postwar additions to the Olivetti campus was Giorgio Soavi, writer, novelist, and poet, no relation of Wanda Soavi. His daughter Albertina recalled that he met Paola in Fiesole in 1948

Albertina Soavi, aged nineteen, in 1971

when he was in his early twenties, just out of the army and starting to publish short stories.

Paola introduced Giorgio to Adriano, and in short order he had his first well-paying job as editor of the Comunità journal. Adriano was aware that his daughter Lidia, she of the porcelain skin, ravishing golden-red locks, and dainty figure, lacked a boyfriend. Adriano introduced them to each other, and rather soon afterward Giorgio had a wife as well.

What no doubt endeared him to Adriano was a parallel gift for seeing others' talents, which, in time, was developed to a fine art. More than one youthful and aspiring talent was pushed along the ladder of success by Giorgio Soavi. Milton Glaser, the American graphics designer, who was one of them, said, "The atmosphere at Olivetti made it difficult to do mediocre work. Soavi had a way of selecting the perfect assignment for each of the designers he worked with and inspiring them to do their best work. He was the best art director I have ever worked with, although he resisted that title, perhaps because he was also a poet, novelist, critic, biographer and maker and collector of extraordinary objects.

"My work with Olivetti, and more precisely with Giorgio Soavi, was one of these golden periods that shed its glow over one's life and influenced it in countless ways, both professionally and personally. . . . He was a man who thoroughly believed in Diaghilev's dictum, 'Astonish me.' "

Soavi also spotted, among many others, Jean-Michel Folon, the Belgian artist, illustrator, and Surrealist, and Paul Davis, American graphics designer, at early stages of their careers. In his long career with Olivetti, Soavi commissioned such articles for Comunità's journal as "The Realism of Courbet" by Giulio Carlo Argan, "Eugenio Montale,

Giorgio Soavi, a portrait
by Graham Sutherland, 1963

or the Poet as Militant" by Geno Pampaloni, and "The Poetry of Mario Luzi" by Franco Fortini. He was soon directing a series of handsome literary classics, illustrated by contemporary artists. From 1959 on, he designed a desk diary that Olivetti published annually for almost four decades. Among the many artists he commissioned, ranging from Balthus to Bacon, was the British artist and controversial portrait painter Graham Sutherland. The latter's depictions of his subject were sometimes more searing than the sitter could endure, as was the case with Sir Winston Churchill. But when he came to paint a portrait of Soavi, Sutherland was admiring, did full justice to his subject's good looks, and the only thing odd about the result is that Soavi is shown tied to a chair, his hands behind his back. That symbolized, Sutherland explained, the fact that every artist is bound by his talent. Or perhaps he meant immobilized.

*A high-style advertisement for new
Olivetti products, 1950s*

That ability to discover new talent eventually led to the breakup of Giorgio's marriage to Lidia. Albertina said, "My father was always looking for talented young artists because Adriano was so interested in expanding the cultural connections. Giorgio heard about this young Sicilian, Bruno Caruso, just starting his career, very poor but very gifted. He met him in Milan and began hiring him. Unfortunately my mother met Bruno as well. I say unfortunately, because at that moment the marriage was not going well. In a short time, she decided to move to Rome and start a new life with Bruno." One assumes that Giorgio took the break with something approaching equanimity because he agreed to take part in a photo shoot that caused a minor sensation in the family at the time. In the portrait Lidia, a leopard fur coat slung over her shoulder, rests a chin lightly in one hand and stares thoughtfully into space. Bruno, in a dark suit, sits on her left.

Giorgio, wearing a similarly neutral suit, stands beside the couple, his drooping head and oddly positioned hands suggesting the helplessness of Jean-Louis Barrault, the mime who never gets the girl in *Les Enfants du Paradis*. The physical likeness between the two husbands is quite striking.

Adriano's rapidly expanding cultural circle was, as might be expected given his fondness for publishing, also peopled by aspiring young authors. Paolo Volponi, writer and politician, whose works reflect the dark side of industrialization and capitalism, worked for more than twenty years at Olivetti, Fiat as well. (He won the Strega Prize, one of Italy's highest literary awards, for *The Worldwide Machine*, about the dangers of automation, in 1965.) Another distinguished author, Ottiero Ottieri, who was in sympathy with Socialism and published a history of that party's first century, was hired by Olivetti in the 1950s as personnel manager, working side by side with Volponi. Others with connections to Olivetti's cultural circles include Franco Fortini, the pseudonym of Franco Lattes, an Italian Marxist essayist, poet, and translator, and Giovanni Giudici, Italian journalist and much admired poet.

It was a stimulating place to work. There was a remarkable feeling of freedom, personal and intellectual, as evinced by the many courses in history, civics, art history, along with concerts, film festivals, and art exhibitions. Olivetti's quest to synthesize political and social goals, along with modern architectural thought, led as before to new theories about town planning. As publisher, a translation of Lewis Mumford's book, *The Culture of Cities*, was high on the list (1954). Other published works include *Italy Builds* by George Kidder Smith (1954), *Architecture and Society* by Erwin Gutkind (1958), and works by Le Corbusier, *La Charte d'Athènes* and *Les Trois Établissements Humains*.

Yet another distinguished former employee, Furio Colombo, author and professor, was starting his journalistic career in Milan when he was discovered by Olivetti. One evening in late 1959 he was leaving the Olivetti offices in Milan when he saw Adriano "wandering about in the parking area." It seems he was on his way to catch a train but did not have a ride so Colombo somehow piled him and his luggage into his modest Fiat Topolino and took him there. Colombo said,

"On the way he told me at once that he had bought Underwood, and would I like to go to the U.S.?" Of course Colombo said yes.

"Among other things, he wanted me to hire workers. Then he said something like this, 'You should choose people like yourself for the electronic project we are working on. They don't just have to be engineers, mathematicians or physicists. They must be intelligent and adventurous, longing for new experiences, with an open-minded attitude, able to dream, think about and finally carry out ideas.'"

This irresistible offer appears to have come quite soon after Olivetti's previous invitation to join him in his political adventure (1958). Colombo had demurred. He said, "I could not imagine taking part in a political campaign linked with an employer-employee relationship. He seemed to think that was no problem at all. I should just put one aside for the other." Adriano continued to press this offer and Colombo to decline, with many reservations. What if he lost his job? "Olivetti by then was an extremely powerful person and I could imagine the consequences." But Olivetti listened kindly without comment. It seemed that Colombo not only retained his job but received a promotion. A couple of days later, "He made me a director!

Outside the main entrance of the via Jervis, 1950s

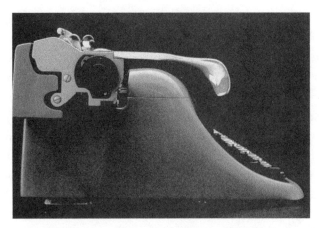

*A profile portrait of one of Marcello Nizzoli's advanced
typewriter designs, the Lexikon 80, 1948*

"It was one of the most instructive lessons of my life. I always
hoped to meet someone like him again. But I never did."

Thanks to some government help and international economic aid
the company had recovered rapidly, and a brisk export trade began.
Employee numbers began to mushroom and, in the years from 1947
to 1951, Olivetti's capital increased by a factor of ten, going from 120
million lire to 1.2 billion. Instead of simply reissuing prewar models
Olivetti reinvented them from the inside out. The company's brilliant
designer, Marcello Nizzoli, received immediate recognition for his
radically redesigned office typewriter, now called the Lexikon 80,
and it evolved, according to the Museum of Modern Art in New
York, into "the most beautiful of the Olivetti machines. The blank
metal envelope in the hands of a sensitive designer has become a
piece of sculpture. . . . Many of the little levers and handles which
one finds grouped at both ends of the carriage . . . and which are
responsible for the bristling look of most models, are . . . ingeniously
bound up with the larger body. By this means a look of order and
simplicity are achieved." The occasion was an exhibition devoted to
the aesthetic qualities of Olivetti's machines, "the leading corporation
in the Western world in the field of design," and the first of many
such exhibitions.

In the same way Olivetti's new portable, the Lettera 22, was not just an improvement on the clunky prewar version (the MP1 in 1935) but equally transformed by Nizzoli's magical touch. Taking his cue from the car industry Nizzoli used steel that could be pressed into curves, hid the key rods, incorporated multiple levers, and produced a similarly admirable balance of sleek, curved, and flat surfaces. The Lettera was lightweight, distinctively packaged, and beautifully engineered. Fairly soon it became a badge of professional pride to legions of peripatetic journalists and was adopted by well-traveled authors as well, from Gore Vidal and Luchino Visconti to Günter Grass (who owned three). A public garden in Milan, named for the celebrated journalist Indro Montanelli, is dedicated not only to him but to his Lettera 22. In 2009 when Cormac McCarthy's Lettera 22 was auctioned at Christie's, it sold for $254,500. The Lettera 22 design has entered the hall of fame dedicated to industrial design excellence by being named one of the hundred best such examples of the century.

Then there was Divisumma, the company's newest version of a calculating machine, the result not just of Nizzoli's superb design but a real advance in calculators themselves. Natalino Cappellaro, who joined Olivetti as a member of the general workforce, rose quickly through the ranks because of his superior ability to rethink mechanical products. At that stage Olivetti's calculators, which sold well, had been constructed of all-mechanical moving parts. Cappellaro went to work and came up with the idea of an electromechanical calculator that also had a printer, a real advance. The first of many versions appeared soon after World War II. It was not only the fastest machine in the world in terms of multiplication but its ability to print up results was a sensation. To build a Divisumma, as it was named, took ten man hours, the same amount of time as a mechanical typewriter, but the profit was huge: three times more than it cost to produce. Alberto Vitale, who joined the company in 1943 as head of the machine division and rapidly became a director, said, "The invention took off like a rocket. Nothing could match it, and Olivetti had a fantastic cash flow." Factories sprang up to keep up with the demand; four of them, in Italy, Spain, Brazil, and Argentina, were dedicated to the production of Divisumma. Eventually six million were sold.

The much admired showroom off the Piazza
San Marco in Venice, 1950s

The Harvard Business School, which published an extensive review
of the Olivetti Company in 1967, observed that in 1953, the year Dr.
Ugo Galassi was appointed as head of sales, the number of employees
began to double and triple again. Prices were competitive—too competitive for some of Olivetti's rivals—product quality was on a par with
excellence of design, and salesmen were everywhere. "Throughout the
world, Olivetti became known for the number of men [Galassi] would

field in the marketplace. One French competitor said: 'The trouble with competing with Olivetti is that, if you throw two of their men out the door, another one will be trying to climb into the window.'" Vitale, who became a very good friend, said, "He was a very good manager, the only one of the managers who could have succeeded Adriano. Whether he had the same vision I don't know, but he had great charisma." Part of Galassi's success had to do with his ability to inspire and refusal to accept less than the best. "If sales were down, he would be in a rage. 'How dare you bring me these numbers?' he would say." Nevertheless, "people would go to the end of the world for him."

Diaghilev's maxim, "Astonish me," became the guiding principle. The showrooms that sprang up in Rome, Milan, Venice, Paris, Barcelona, and elsewhere were imaginative examples of what can be done to entice the public to buy. The company experimented with murals, installing an early example in their Rome showrooms. As the decade of the 1950s progressed the designs became more elaborate and ambitious and, in many cases, specifically focused on what could be seen from the street and not just inside. The objects themselves, with their unpromising boxy shapes and neutral colors, had to be presented as if they were enticingly mysterious, inviting the passerby to investigate them in a grown-up, magic toyshop. Light, space, color, movement, trick reflection: everything that could be done was done, the more unexpected the better.

In Paris, on the rue du Faubourg Saint-Honoré, tubular constructions descended from high ceilings to meet phalanxes of machines on circular platforms lit from below. Display units became ascending staircases of white plastic that disappeared into space. Paintings by Klee and Chagall graced the walls and strange, semi-abstract figures reminiscent of Giacometti stalked the interior. The result was so striking that the showroom began to serve as an impromptu gallery space.

The Olivetti showcase on the Piazza San Marco in Venice is cited as a particularly high example of this increasingly sophisticated mingling of art and sculpture with sleekly designed objects for sale. What began

*The huge sculpture by Costantino Nivola made from
sand-casting that graced the Fifth Avenue showroom
of Olivetti office machines in the 1960s*

as a long, dark, unpromising space, in the vision of Carlo Scarpa, an
Italian sculptor, became a mysterious and seductive inner sanctum of
marble, Murano glass, and mosaic tile. Scarpa's solution was to design
a staircase miraculously built of individual marble slabs supported
by rods that seems to float weightlessly in space. At the entrance
Scarpa designed an abstract sculpture, *Nudo al Sole,* mounted on a
black marble basin that trickled with water. His design was so much
admired that, after the Olivetti showroom had closed down, it was
rediscovered, restored, and is now open as a museum.

Perhaps the flagship that gave Adriano Olivetti the most pride was at Fifth Avenue and Forty-Eighth Street. When it opened in this prime location he celebrated the event by giving a speech. Something about a cherished ambition that had finally been met. Great care was taken to put the best of Italy on display; pink and green marble, Murano glass lamps hanging on long wires from the ceiling—no expense was spared to turn a visit into a discovery of marvels. Among them was a justly famous mural designed by Costantino Nivola, from a Sardinian family of stone masons who had handed down the complicated technique of creating sculptures from sand-casting. The lengthy process involved designing a reverse image in sand which would become the mold for a cement casting. Once the cement had dried the sculpture was installed at Forty-Eighth and Fifth to great acclaim. After the showroom closed in 1970 and its design was dismantled, the cement mural went to the Science Center of Harvard University, where it has been ever since.

Perhaps the best known aspect of the New York showroom was the Lettera 25 typewriter that was placed just outside the door, bolted to a free-form green marble base. The general public was invited to try out the machine. Try it out they did. Ten months after the typewriter was put in place, in May 1954, it was determined that fifty thousand people had stopped to try out the machine, many times replaced, since mischievous children did a lot of damage. Some of the visitors were regulars like the poet Frank O'Hara, who liked to tap out a few verses while, he said, "strolling through the noisy splintered glare of a Manhattan noon." Some were gleeful: "I'd give up my spaghetti for this here Olivetti," one girl wrote. Many took refuge in the time-honored maxim, "Now is the time for all good men to come to the aid of the party," invented by Charles Weller, a court reporter, to test the speed and efficiency of his new machine. (A slightly better test is "The quick brown fox jumps over the lazy dog.")

Some wit calling himself Marlon Brando offered the world a blanket invitation to visit him any time and posted a "Hollywood" number, 5-6000. How many people tried is unknown. Perhaps the best image (passersby were being secretly photographed by a *Life* magazine photographer) was that of one mysterious visitor, well camouflaged with glasses, a large, wide-brimmed hat, and a capacious raincoat. He

Lidia and Roberto in Rome, c. 1950

wrote, "jolhajoyrbakjg:nn." Wait, was that a coded message? Well, maybe not. He added, "I can't type."

While Adriano struggled with the herculean task of keeping everything going at once, with export sales rising seven times for Olivetti typewriters and twenty-three times for calculating machines, his heir apparent was growing up. The boy with blue eyes and curly red hair, who looked just like Camillo and charmed everyone, took an undergraduate degree in statistical analysis from the Luigi Bocconi Commercial University in Milan and graduated in 1952. Eight years earlier, in 1944, when he was sixteen, he had been allowed to move into his own apartment and shared it with Franco Ferrarotti. "The apartment in Milan was near the Piazza della Scala and I got to know him well. We would go out together," Ferrarotti recalled. "He had a good friend in a statistics professor. But I thought Roberto at that age was too frivolous, over-indulged by his parents. Too many people making his life easy. He would go up to the mountains to ski, his engine would get frozen and someone would always go and get him." Since everyone knew who he was, he had no trouble finding a girl and there were always too many, which Roberto enjoyed. But then, Roberto would want to break up and would try, in his father's time-honored fashion, to pass the girl along to someone else. But this was not yet the case with his short-lived first marriage to Vittoria Beria, who happened to be the twin sister of his close friend Ricardo. Vittoria was "stunningly beautiful," the kind of person who turned heads in a restaurant, said Milton Gendel, an American expatriate writer-photographer, who fell in love with her too. When that did not work out either, Adriano, who had introduced Vittoria to him, saved the day by making her his hostess when he was in Rome.

Ferrarotti continued, "Roberto had a wonderful relationship with a brother-in-law of Adriano's, Gino [Levi] Martinoli, who worked as a manager at Olivetti's and they went everywhere together. They were always trying to get me to learn how to ski," but Ferrarotti couldn't see the point. He did not mention that Roberto was already something of a menace, with a passion for speed, whether racing down a mountain, in a boat, or down Central Park West in a Buick convertible—like most Olivettis, he became fluent in English and, as the joke went, could flirt in German. One of his first accidents involved his friend Pier Mario in Ivrea. They had a bit too much wine, went out boating, and managed to capsize the boat. It was perfectly amazing how many accidents Roberto managed to have without fatal consequences. David Olivetti said, "Adriano once told me that if he had counted up all the scrapes he had been obliged to get Roberto out of, he could have filled three handwritten pages. They were mostly minor, often just stupid things, like trying to change a sweater while driving and hitting a wall," he said.

Although Roberto was taking the right courses, in all the public and well-documented appearances Adriano made, the lectures, presentations, receptions, and trips, Adriano on holiday, or Adriano touring a factory, only one photograph has come to light showing the two together. This bears out the frequent observation that Adriano did not seem to be grooming anyone to replace him and particularly not Roberto. A certain pattern begins to appear. Camillo, who grew up without a father, assuming a takeover role as he matured, began to treat his mother as someone who needed to be taught how to lick a stamp, ran his business as a one-man adventure, and assumed the role of benevolent patron at an early stage. His wife, a quiet confidante, was always very much in the background. And the first bicycle he ever bought had no brakes.

One does not see the same paternal passion for speed in Adriano. He drove more like the absent-minded professor who never remembers to shift gears. But neither does he seem to have been much of a father either. How much he ever saw of Camillo, who would be called a "workaholic" nowadays, is a moot point. Circumstances also worked against any relationship between father and son in Roberto's case. His parents separated when he was still small, the war intervened, his father had to flee, and for six important years, when Roberto went from puberty to young manhood, he had no father. Desire, Roberto's daughter, thought her father was in awe of Adriano and cowed by him at the same time. "Roberto would suggest something and Adriano would dismiss it; 'we tried that and it didn't work.'"

She added, "I imagine that when he was with Adriano he always had to be 'in control.' For the job, for the workers, for the family, and for the future. He was brought up strictly, expected to be good, always to act properly, use the right protocol, lots of ethical training, acting honorably and so on, always thinking of others." He was hedged around with obligations.

He was a stoic when it came to physical pain. Once when they were skiing together, he was wearing knee britches and thick red woollen socks. As they were climbing one late afternoon, Desire noticed with alarm that one of her father's socks was red for another reason: it was covered in blood. They stopped. "Oh yes," Roberto said. He pulled down the sock to discover a splinter of wood embedded in his thigh. He ripped it out, pulled up the sock, "and we went on skiing." He was, everyone said, very courageous but sometimes it was hard to tell whether he was being brave, or merely foolhardy. Like his father, he certainly took a lot of risks. The photographer Inge Feltrinelli Schoenthal, an old friend, recalled that she and her late husband, Giangiacomo Feltrinelli, owned a Piedmontese country house, something like a castle, that was semiderelict. It was just forty minutes from Ivrea; Roberto loved it and went there often.

"Once, in the middle of winter—between Christmas and New Year's—he visited our sauna and then the lunatic jumped into our indoor pool which (being unheated) was covered with ice, and began swimming as calmly as any Finn." She was clearly torn between admiration and exasperation. She recalled another time when he set off for Sardinia all by himself, in a rubber dinghy, and got there, too. You never knew whether, on the beach at Forte dei Marmi in fine weather, Roberto might decide to go for a dip, take off his shoes, and wade calmly into the water fully clothed in his best white linen suit. Like skiing, boating was his release. There is a smiling photograph of him seated on the deck of his boat, bare-chested, and with bare feet. Desire said, "He raised me as a boy to think nothing was dangerous. 'Just go through it,' he said. The only thing he would not do was a parachute jump."

Everyone commented on the close bond between Roberto and his mother. What is interesting about the photograph that was taken of them in the Fiesole garden is the touching way he tries to protect her, hurt and defiant. Their poses mirror each other, as do their clothes, and the interesting detail is the identical watch both are wearing, rectangular with a black band. People mention the fact that he himself designed a ladies' watch, a slimline rectangular version with a plain black band. He had two copies made and often wore one of them, which looked very much out of scale. Desire said, "Imagine that small watch on his big bony wrist!"

The cult of the mother-son bond in Italy is too well known to need emphasis, and the particular bond of the Jewish mother with her son is a perennial source of jokes, most of which revolve around a Jewish mother's willingness to jump off a cliff if it will make him feel better. As John Hooper observes in *The Italians*, "it would be hard to come up with a more forceful, overprotective parent than the stereotypical Jewish mother." "Mamma," a popular song from the 1940s which was revived by Luciano Pavarotti, concludes with "These words of love that my heart sighs / Are maybe not used these days / Mamma! / But my most beautiful song is you / You are life / And for the rest of my life I shall never leave you again." Hooper goes on to state that, in the English lyrics for the same song, the sentiment is limited to a longing to see Mamma again rather than words that sound a bit too incestuous for comfort. And it is sung by a girl.

As for Paola, Desire recalled, "When I was small I was her sweetie and she was mine. She loved me because I was Roberto's beloved daughter." She was also tolerant of the many nephews and nieces who joined her for a seaside holiday every summer at Forte dei Marmi, where the Olivettis kept a villa. Desire recalls that one of the boys wanted to ride his bicycle in the waves. Rather than tell him it would not work, Paola conducted him and his bike to the shore and let him find out for himself. She was, of course, a meticulous hostess, with a surprising knowledge of household maintenance. "She knew everything about fashion, including how to sew, how to repair. She could tell you the styles for art, for fashion, for everything, how to take care of furniture, how to clean the silver. . . . She put passion into everything. I remember seeing her wax the furniture because she felt the pieces needed nourishment. Everything had to be cared for. The garden, everything. You had to be properly dressed, behave properly. She would not leave the house without a hat. Every morning. Buffing nails was very big. Every morning, a half an hour in front of the mirror to do her face. She would put on make-up before she opened the door to the postman."

After a dinner party, if the servants had gone to bed by then, Paola would scrape the plates and assemble everything in a meticulous pile for the maid to deal with in the morning. She would cover everything with a cloth and call it her "hypocritical" pile, a play on the word "hypercritical." She thought that was a great joke. She would then retire to her bedroom, her portrait of Proust, and her pile of books to read. "She was a very romantic person, very profound, with very deep thoughts," her grand-daughter said loyally. Quite what Roberto's successive wives felt about her—he married Anna Nogara and Elisa Bucci-Casari after Vittoria—is not clear. One wife thought that Paola was "too protective" of Roberto. Anna Nogara, Roberto's second wife and Desire's mother, while appreciating her special qualities, said that Paola could be hard to get along with. "One time I asked Roberto, 'Why is Paola so difficult?' 'Because she is,' he said. 'Because she is.'"

—

People liked to think that Paola was ready to get back with Adriano, if asked, or he was ready to get back with her if she asked him, but neither would speak first. That may or may not be true. But sometime after World War II someone new took possession of his affections, a girl he had dreamed about since he was a small boy, the fairy in *Pinocchio* with blond hair and blue eyes. Her name was Grazia, the daughter of Paolo and Germana Galletti. She was born in November 1925 in Piacenza but since her father was a colonel in the army the family moved often. By the time World War II began the parents appear to have separated. Paolo was posted to Africa and Germana and her daughter moved to Rome. Beniamino de' Liguori Carino, one of her grandsons, who now heads the Fondazione Adriano Olivetti in Rome, said of his grandmother, "Germana was a most beautiful and elegant woman, a professional pianist who gave piano lessons and also worked in embroidery, specializing in clothes for children's boutiques." Grazia recalled that the bombardment of Rome started one day when she was on the streets, on her way to sell some embroidery that she and her mother had just finished. They fled to Ivrea and then Turin, dodging SS patrols, and then returned to Ivrea. Grazia and Adriano might have first met when she was a teenager. In any case, that they should meet at some point was inevitable. Wanda Soavi was Grazia's aunt.

Adriano's deep attachment to Grazia is something of a mystery and perhaps best explained by unresolved emotions from adolescence. Writing in 1948–49, when Grazia is in her early twenties, he explained that he was afraid of losing her and this made him feel stupid and "gauche," like an awkward adolescent. In those days an age difference of twenty-five years was considered more of a handicap to marriage than it is nowadays, and the reaction of her Catholic parents, even separated ones, can be predicted. But he insisted he wanted to marry her, and why couldn't he? He could remove the barriers. The comment was made when divorce was still outlawed in Italy (it did not become possible until 1970) but there were ways around that and indeed Adriano had succeeded in having his marriage to Paola annulled in

San Marino. He was willing to be baptized as a Catholic. Anything, as long as he could win the blue-eyed, fair-haired girl of his dreams.

Quite how the relationship ended with Wanda is unclear. A story told about that is that Adriano rang up Wanda one day and said, "Let's get married tomorrow! I insist!" and she demurred, "But Adriano, I don't have anything to wear!" At which point Adriano banged the receiver down and married Grazia instead. Something is clearly missing here. Some kind of conversation could only have taken place already and the likelihood is that Wanda had been insisting on marriage and now she was throwing up objections and he was exasperated. And what about her feelings when she discovered that, instead of her, Adriano was marrying her niece? It sounds like the plot of a third-rate Italian opera.

Adriano Olivetti and Grazia Galletti were married in the Church of Santa Prisca all'Aventino in Rome in January 1950. The groom's children, who were about the bride's age, found reasons to be elsewhere. Roberto was somewhere in Austria, and Lidia was off skiing with Giorgio Soavi, her future husband, in Sestriere. David Olivetti's handsome bound volume of the Olivettis, which is part family tree, part a record of weddings and births, has no photographs of the Adriano-Grazia wedding. But there are several of the proud parents at the end of 1950 holding the new arrival, with Laura, or "Lalla" in their arms.

In this letter to Grazia, Adriano goes into some detail about the religious sense of mission that lay behind everything he was doing. His whole life, he wrote, was dedicated to building a foundation for God's reign, even if the stone he placed there was only one of thousands of such stones. God had come to him with the example of the rich young man who gave everything he owned to the poor. He loved the poor, those who suffered, who lacked even a blanket to sit on or a coat to wear. "[T]he redemption from misery and the fight against egotism is my life, even if I know that my path is long and difficult. God is good and generous to me and I know he will assist me. I will not betray his trust."

—

Just before Laura was born, on December 6, 1950, Grazia came back from a visit to the clinic one day to find her husband sweating and gasping for breath. He was having a heart attack and it took him several months to recover. And no wonder, because in 1950 the company was expanding at a pace the Harvard Graduate School of Business Administration called "feverish," requiring superhuman energy and concentration. He had shortened the work week, a standard forty-eight hours in those days, to forty-five hours, which was considered unprecedented, adding further benefits and privileges that solidified the company's lead in worker contracts. A skilled workforce, well paid and cared for, would prosper and so did the company. During the late 1950s and early 1960s, Olivetti's share of the world market for typewriters continued to leap ahead until it was close to 27 percent and even better for calculators: 38 percent. By the 1960s Olivetti was one of the world's leading producers of business machines, the sixth largest industrial company in Italy, and No. 103 on *Fortune* magazine's list of 200 largest industrial companies outside the U.S. It operated in 117 countries with 54,000 employees around the world. Its consolidated sales of approximately $450 million were split between a domestic market share of 20 percent and 80 percent abroad.

The company's physical expansion was equally rapid. A new, improved employee canteen in a hexagonal design (a tribute to the ideas of Frank Lloyd Wright) replaced an earlier circular building, and construction began in 1953. The following year ground was broken for an entirely new Social Services Center on the via Jervis. A study and research center was already in construction (1951–). Perhaps the most ambitious project, designed by Figini and Pollini, was a huge expansion of the main factory that began in 1955 and continued for several years. Another nursery school, factories in Argentina and Brazil, a summer holiday resort for employees at Brusson in the Ayas valley (1960–64)—the list went on and on. And then there was Matera.

Carlo Levi's depiction of peasant life in *Christ Stopped at Eboli,* published in 1945, not only described the desperate situation of the local peasantry but also those of the nearby town of Matera. He also visited an ancient cave community called the Sassi (Italian for "stones"), which had been inhabited since Paleolithic times. In the 1930s it was inhabited by roughly sixteen thousand of the world's poorest; illiterate, suffering from malnutrition, malaria, trachoma, and even anthrax, who were existing in a tangle of buildings, caves, and rocks that honeycombed the rocky flanks of a steep ravine. In those days it was called "the shame of Italy." Levi described a hallucinogenic aura of decay, "like a schoolboy's idea of Dante's Inferno," he observed. What had been prehistoric cave dwellings had become filthy and disease-ridden, "where barnyard animals were kept in dank corners, chickens ran across the dining room tables, and infant mortality rates were horrendous."

After the prime minister, Alcide De Gasperi, toured the Sassi in 1950 and pronounced it "a national disgrace," the Italian government joined forces with the United Nations to construct a new village, La Martella, and move thousands of people out. Olivetti, which was committed to the development of the south, took over the major role of planning, which would replace the hovels with new houses offering water, light, heat, and indoor plumbing. It was a rush project for political reasons, completed in just three years and at the cost of cheaply built houses with persistent problems that included damp. The irony is that the Sassi, architecturally complex and replete with evidence of prior inhabitants from earlier centuries, also contained many treasures, such as priceless Byzantine frescoes in the old churches made of rock. These have been rediscovered and turned into chic hotels, restaurants, art galleries, and the like. Some Matera families have revisited the ruins, cleaned them up, added modern conveniences, and put them back as bed-and-breakfast rentals on Airbnb. Tourists can now sleep where shepherds once housed their flocks and chickens ran across the dining room tables. So the world turns.

Feeding, housing, and educating the poor would never be enough for Olivetti, no matter how laudable such goals might be. As always, he was intent upon reconciling the assembly line with the human beings condemned to serve its demands, an issue he had struggled with ever since he first recoiled in horror. The project that came closest to doing this was his complex of factory and buildings in Pozzuoli.

The town of Pozzuoli on the outskirts of Naples is, like Matera, many-layered with the evidence of its long history. The Greeks had a colony there; when the Romans arrived in 194 BC they gave it the name of Puteoli, having its roots in "puteus," a well (and overtones of sibyls predicting the future) and "puteo," the Latin for "stink," a reference to the sulphurous fumes from its volcanic past. In ancient times it was a great port, and a Roman naval fleet based nearby was the largest in the ancient world.

In recent times Pozzuoli became very poor, its economy largely dependent on tourism and agriculture. Adriano wanted to change all that. "He was making a pilot experiment in a depressed area," *Fortune* magazine wrote. "He chose a poor town with little employment, yet not so poor as to be crushed. His aim was to found a productive enterprise in the presumably inefficient south and lure other industry into the area. . . . Most of all Adriano and his architect, Luigi Cosenza, had selected a glorious site overlooking the Gulf of Baia, with Capri and Ischia shimmering in the distance. They wanted to create a workplace that could be "an instrument of fulfilment and not a source of suffering," Adriano explained. There would be no windowless walls or air-conditioning. Glass would replace bricks and cement, and ingenious air ducts, the necessity of artificial cooling. The emphasis would be on the beautiful surroundings so that trees and grass, birds and flowers were a constant "source of comfort in the working day." Nature would thread itself through the daily round instead of being barred from it.

Cosenza's design is considered a model of what can be done to take maximum advantage of a sloping site with panoramic views. He said, "I wanted to make the factory less like a prison. You can't change the fact that it is a prison, but I wanted to make it as different from the old-fashioned factory as I could." He succeeded. The result was a building "that might be taken for an elegant resort hotel or a sanatorium of the modern style. . . . Cosenza has created a plant which is unique in its sense of freedom and elegance," *Fortune* wrote. "From every bench and point of the assembly line there is a view of the outdoors. Light pours in with all its Mediterranean clarity and abundance. Spotted here and there among the low buildings which are joined together in a coherent whole are ponds, caught in free-form concrete walls. Landscaping encloses the entire industrial bulk and reunites it with nature."

The Pozzuoli complex would also include a canteen overlooking the Bay of Naples, a social services center, a training center, a library, church, stores, and schools. It was the genius of a company, the famed *Tocco Olivetti,* once again turning aesthetics into profits. For those lucky enough to work there the Olivetti plant represented an impossible dream come true. But for Adriano the Pozzuoli experiment was part of a larger objective, the lesson being that pleasure in work results in the creation of beautiful objects. He might well have echoed the final thoughts of William Morris, who concluded in *News from Nowhere* sixty years before (1890) that, "If others can see it as I have seen it, then it may be called a vision rather than a dream."

The Experience of a Lifetime

"And it is all one to me / Where I am to begin; / For I shall return there again."

—Parmenides, Elean philosopher

J oseph Stalin died in March 1953 and was replaced by Nikita Khrushchev. That, by coincidence, was the month that a new ambassador from the United States arrived in Rome and the change of personalities at the highest level of the Communist world would have made very little difference to the new appointee. She was the first woman to have been awarded a major diplomatic post and, no matter who was in power, she was a fervent anti-Communist.

She was Clare Boothe Luce, politician, playwright, war correspondent, a charismatic speaker, and the wife of Henry Luce, publisher of *Time, Life,* and *Fortune.* She was also known for her barbed comedy, *The Women,* 1936, which became a Broadway hit. She was, incidentally, adept at wry epigrams that have found their way into common parlance, such as "No good deed goes unpunished" and "A hospital is no place to be when you're sick." She was determined to be a success because, she also noted, if a woman is a failure, people argue no woman can succeed and she intended to prove them wrong.

In the five years since the Italian elections were expertly and expensively manipulated to keep Communists out of office in Italy, little had changed. Italy was still considered the frontier of the Cold War, marking "the dividing line between contesting ideologies," Daniele

Ganser observed in *NATO's Secret Armies.* "On the left side of this frontier stood the exceptionally popular and strong PCI Communist Party, supported with secret funds from the Soviet Union. . . . On the right side of the frontier operated the CIA and the Italian military secret service with its Gladio army and a number of right-wing terrorists." Among them were not only former Fascists like the Black Prince but members of the Mafia. The Mafia boss Charles "Lucky" Luciano, who was serving a fifty-year prison sentence in the U.S., had been released because he could work with the Sicilian Mafiosi to facilitate the Allied landings of that island in 1943. After the victory, the CIA was happy to "maintain a clandestine friendship with the Sicilian Mafia in the name of combating Communism in Italy," and that included the Gladio forces. Following the Russian detonation of an atomic bomb, the launching of its spaceship Sputnik, imposing the Berlin Blockade of 1948, and the partitioning of Germany,

*The new ambassador, Clare Boothe Luce, meets
the welcoming committee in Italy, 1953.*

tensions between the great powers were, if possible, more acute than before. Such a state of affairs was further exacerbated by the arrival of a certain senator from Wisconsin, Joseph McCarthy, who claimed that traitors—Communist sympathizers—infested every aspect of American life.

There was particular reason for concern in Italy just then. In the new elections of June 1953, the leftist coalition of Communists, which included the Socialist Party, had increased in strength. William Colby, new director for the CIA in Italy, who arrived that year, was concerned that "the combined Communist and Socialist vote would grow to become the largest political force in Italy." "My job . . . was to prevent Italy from being taken over by the Communists in the next—1958—elections and thus prevent the NATO military defenses from being circumvented . . . by a subversive fifth column."

The era of CIA interventions was hardly limited to Italy. Some well-documented studies, in particular William Blum's book *Killing Hope: U.S. Military and CIA Interventions Since World War II,* listed more than fifty attempts to assassinate politicians, beginning with Kim Koo, Korean opposition leader in 1949 and followed by Mohammad Mossadegh, prime minister of Iran in 1953. These continued into the 1960s, including the bumbling attempts to end the life of Fidel Castro in preposterous ways, including an exploding cigar, until the Church Committee put a stop to such "targeted killings" in 1975.

In Italy, the manipulation of votes was partly being accomplished with sizable bribes, or what the new ambassador called "arm-twisting," disguised as foreign aid. Besides the Marshall Plan, Congress had approved a program that awarded contracts to NATO countries to manufacture components of military hardware, called "Offshore Procurement." But as Ambassador Wells Stabler, who was a political officer in Rome at the time, stated, the contract could only be awarded if the trade unions involved voted democratic. If the factory were primarily Communist, there would be no contract. Stabler said, "There was a lot of hard ball played over this, as it was felt that this was a major element to try to block the Communists."

In addition to cars Fiat also produced military hardware and was extremely anxious to acquire "Offshore Procurement" contracts. The negotiations were being handled by Vittorio Valletta, whom Ambassador Luce, in her reminiscences, called "the brains of the particular thing," before Gianni Agnelli, head of the family who owned Fiat, took charge. Asked if Valletta, always called "Il Avvocato," was as clever as the well-known Lee Iacocca, once CEO of Chrysler, the ambassador replied that Valletta "could have run rings around" Iacocca. Valletta was constantly in her office trying to negotiate the coveted contract. After each visit she would receive an enormous box of red roses.

Gianni Agnelli, chairman of the Fiat company (undated, c. 1960s)

Her oral history continued, "The day came when Gianni Agnelli himself, together with Valletta, came on a very big contract." The problem was that, in the June elections, the Fiat trade unions had voted Communist. So she had to turn him down. Gianni Agnelli pleaded with her to change her mind. Valletta begged her to do it. "And I said, 'No.' I mean, they were there for two hours." She continued, "Agnelli left first, and Valletta stood at the door, talking to me for a minute, and then Gianni Agnelli yelled at him and he went along out." That night she received the biggest bunch of roses she had ever seen and they were exquisite. "I thought, 'My God, I know what happened. He ordered them before this conversation.'" He was much too sure of himself. The little "Avvocato" denied it later.

Ambassador Luce had good reason to be proud of having stared down two of the most important industrialists in Italy. As Thomas W. Fina, who also served in Rome in those years, said, "[W]e had resources at that time, financial resources, political resources, friends, the ability to blackmail, all the things that a great power . . . traditionally has done in dealing with its friends and enemies. We were placing military contracts, denying military contracts, subsidizing political parties, withdrawing money from political parties, giving money to individual politicians, not giving it to other politicians, subsidizing the

*Enrico Cuccia: "a stubborn
and vengeful figure"*

publication of books, the content of radio programs, subsidizing newspapers, subsidizing journalists, granting and denying visas. All of the things both of a covert and an overt nature that a great power . . . in the tradition of what the Fascists, the Communists, the Nazis, the British and the French, had done before, we were doing in Italy."

Fina continued that U.S. diplomats were not allowed to speak to members of Communist or Socialist parties under any circumstances. They were "off limits," and if one met them at a cocktail party one had to look the other way. This, of course, did not apply to the politicians who counted, but these were few and far between. He said, "Italy is an astonishing country in which a very small number of people make the decisions. They all know each other, half of them are related to each other, and even though they are political adversaries, they may be personal friends." And they are all in Rome. "No one would dream of leaving Rome, which is where political intrigue boils from morning to night. You know, if you turn your back, you've had it."

Among the very small number who really mattered in those days was a short, stooped figure who wore gray suits, who had not spoken to the press for decades, and made it a rule never to have his photograph taken. He was Enrico Cuccia, then about fifty, who, as will be seen, played a pivotal role in the fate of the Olivetti company. He was born in Rome in 1907 to a family of Sicilian origins, began his banking career at Italy's central bank, and was selected by the Fascist government to run the IRI, a state holding company set up to manage state property. That was in 1932. Once postwar reconstruction began in 1946 Cuccia took charge of Mediobanca, a fund created to lend money to industry, something that was then banned to Italian commercial banks. Shortly after that, Mediobanca began to buy minority shares in major companies like Fiat and Pirelli and the real ascent of Cuccia began, making him "*the* arc of the Italian network of private sector power," Alan Friedman wrote in *Agnelli, Fiat and the Network of Italian Power.*

"Puppet-master of Italian finance, a stubborn and vengeful figure with a lifelong reputation for making and breaking companies, deals, and even men themselves, Cuccia, as they say in Italy, does not confide even in his own shadow," Friedman wrote. "Paradoxically, Mediobanca became the tool of the private entrepreneurs even though it was controlled by the state. Cuccia used his privileged position at Mediobanca . . . to weave a spider's web of industrial cross-holdings." Under this ingenious system, capitalism in Italy became the province of "a tiny elite group of allies who held key minority positions in each others' companies." Often no money ever changed hands. "The

Giuseppe Battegazzorre, Ivrea's police chief for more than twenty years, at the city's train station with Vittoria Valletta, a former chairman of Fiat and Agnelli's right-hand man, c. 1960

ownership of companies may be shunted around, from one part of the network to another," to the detriment of the ordinary investor, "shifting the contents of one trouser pocket to another."

Such enormous financial advantage meant influence in all kinds of direct and indirect ways. One Milanese entrepreneur who had witnessed just what this meant in his own case said, "They will cancel you from the map if you displease them."

It had been suspected that some allies in this tight little arrangement were more important than others and that Cuccia had long favored the interests of companies like Fiat and Pirelli over others in the group, never mind the public interest that Mediobanca was supposedly designed to serve. Where business and personal interests lock and reinforce each other is always a moot point in Italy; Cuccia and the Agnelli family had been on close personal terms for decades. Alan Friedman wrote, "In the Mafia, say [Agnelli's] harshest critics, every godfather has his *consigliere,* or advisor. In the network of Italian financial power, Gianni Agnelli has Enrico Cuccia."

❦

Somewhere in the verdant countryside around Pisa in Italy was a pretty villa, half hidden by the San Rossore pine forests, just outside the village of Barbaricini. The area is well known for its horseback riding and racing stables and shot to prominence because one of the world's champions, Ribot, had trained and was in residence there. It was said that Ribot could sometimes be seen cantering past the villa in the company of his bow-legged jockey. If by chance you were in the area, either walking or driving, you would have to use every effort not to offend the skittish sensibilities of this most beautiful and valuable of creatures. Or so it was said.

One morning in April 1957, Pier Giorgio Perotto, a young engineer, drove out from Turin where he had spent two years working for Fiat in his Fiat 600, not quite the world's smallest car or, as he said, "all that was left" of that experience. He was about to join Olivetti's new electronics laboratory and would become a leading figure in its race to produce the first European computer. He had undergone a rigorous round of interviews and was amazed to be hired. Arriving at his destination he discovered an elegant nineteenth-century villa with terraces surrounded by gardens, and nothing to indicate that it was actually a business site. The contrast between it and the bleak factories he had left behind could not have been more marked. Perotto wrote, "Not only that, but the . . . director, Mario Tchou, who was the son of a Chinese diplomat, gave the impression of combining high technology with a thousand-year-old culture."

The laboratory was Roberto Olivetti's idea. After graduating in 1954 he spent a year studying business administration at Harvard University. The story goes that he took his father to lunch in 1954 with Enrico Fermi, the Italian physicist and creator of the world's first nuclear reactor, while on a visit to Rome. The reason for the meeting was to convince Adriano to invest in research and development in the promising new field of electronics. The date given is 1954 but was probably earlier, since that year Fermi was ill with stomach cancer and died in Chicago in November. In any event a small research center had already been established in 1952 in New Canaan, Connecticut, by Dino Olivetti, who was now heading up the North American division of the company.

What seems plausible is the discovery of a brilliant young engineer that took place more or less simultaneously with the company's decision to establish a larger laboratory in Italy in a collaboration, destined to be brief, with the University of Pisa physics department. Mario Tchou, tall, and immaculately dressed, was already something of an internationalist. His father, Tchou Yin, had been sent to Rome as Ambassador to the Vatican while Chiang Kai-shek was in power, and Mario grew up there, speaking flawless Italian as well as fluent English. He won a scholarship to study in the United States, and met and married a Chinese girl student in New York. By 1952 he was an assistant professor at Columbia working in the field of electrical engineering and a director of the Marcellus Hartley Laboratory.

Whether Dino discovered Tchou, or Roberto, or Adriano himself, is unclear. The fact is that the project, along with the perfect person to run it, came together in 1955, and one by one, dozens of promising young talents were appearing at the door of the quiet, shadowy villa in the woods. They were, to put it kindly, a scruffy-looking bunch. Martin Friedman became Perotto's immediate supervisor. He was "a likeable British engineer . . . who usually wore a tattered T-shirt with trousers tied at the waist with string." Giuseppe Calogero, who was also in the group, recalled meeting a good-natured Canadian named Webb, "a flower child before there were flower children," who drove around in a car two decades old with no roof, looking like "a bathtub on wheels." People arrived on anything that would move, including broken-down cars with missing doors, rusty bicycles, even on foot. Hardly any of them had much experience with electronics. Remo Galletti, an engineer from Trieste, was treated with great respect because he knew more than anyone else. They came from all over Europe and were uniformly bright-eyed and open-minded in obedience to Adriano Olivetti's insistence that assets of personality accounted for more than experience—which, in any case, was going to be minimal. They were fascinated, tireless, and mischievous. Friedman, the Canadian, was among the worst offenders. Calogero recalled in his memoir that Friedman's idea of fun was to make an early machine look as if it was about to explode. He would hide away somewhere, rig up a long tube attached to the machine and begin to puff smoke through it. He would then laugh helplessly as the others, in a panic, turned off everything and would only belatedly reveal he had done it all with cigarettes.

Perotto's first assignment was to build a console for the new project, called the Elea 9003, their first computer. It was a "mainframe," so-called because computer circuits were assembled on large metal frames. They also used thousands of vacuum tubes and required raised floors to accommodate connecting cables, along with massive amounts of energy and air-conditioning. They were always enormous—memory alone famously required space the size of a tennis court—and designed to solve a plebeian range of problems, such as artillery speeds or code breaking. They were so cumbersome, not to say complicated, that they required their own operators and stayed in situ. One rented them out, rather than bought one outright. During World War II, elaborate analog computers, along with IBM's infamous punch card computers, were used to build the atomic bombs dropped on Hiroshima and

Nagasaki in 1945. But the first real advance came in 1946 with the University of Pennsylvania's ENIAC (Electronic Numerical Integrator and Computer), which was still hugely expensive ($6 million in today's money), larger and heavier (sixty thousand pounds), but achieved in just thirty seconds calculations that once took twelve hours with a hand calculator.

Once the war in Korea began in 1950, new weapons were needed and many millions of dollars in taxpayer funds went into the development of systems that would launch weapons of mass destruction. IBM, which put its services at the disposal of the U.S. government in 1940, had, by 1950, become another arm of the military-industrial complex. One of its missions was to design a computer to be used for an air defense program ordered by the U.S. Air Force, called SAGE. IBM would build fifty-six such computers at a cost of $30 million each. It would also work on launching and tracking satellites for lunar and space shuttle flights. And it would design and develop intercontinental missiles, ICBMs, to carry nuclear warheads. Somewhere between 1958 and 1961—the dates are unclear—the U.S. installed thirty Jupiter missiles at ten sites in Italy, aimed at Russia. The installation was top secret and the actual cause of the Cuban Missile Crisis of 1962 although this was not revealed to the American public. The Italians did not know either but somehow got wind of it and a protest was brewing. It was hard to keep a secret for long, since in Europe at large, some seven thousand missiles would be installed. As if that arsenal of deadly weapons were not enough, within ten years after the end of World War II, both the United States and Russia had developed a hydrogen bomb, which was a thousand times more deadly than the bombs dropped on London, Berlin, and Cologne.

In his book, *My Journey at the Nuclear Brink,* William J. Perry, a former U.S. secretary of defense, writes about the fear of nuclear annihilation during the Cold War that had "unleashed the billions of federal dollars that supported the secret defense work that began in Silicon Valley . . ." In a review of the book, Jerry Brown wrote that Perry knew the ways in which the various pressure groups, from "technical innovation, private profit and tax dollars, civilian gadgetry and weapons of mass destruction, satellite technology, computers, and ever-expanding surveillance," contributed to an arms race that was gaining increasing momentum. He also relates the untold story of a team, assembled by Allen Dulles in 1959, on the question of a "missile gap." Congress was seized by the idea that Russia had more missiles than the U.S., and Dulles's directive was to discover whether this was actually true. It wasn't, Perry said, but that did not stop the politicians. He was also in a select group of analysts put together by President Kennedy during the Cuban Missile Crisis to determine just how dangerous the Russian missiles were. Brown wrote, "Each day as he went to the analysis center, he thought to himself that this would be 'my last day on earth.' " Meanwhile IBM was making rapid advances in computer design at its research center in the future Silicon Valley. Its importance in terms of America's ability to develop advanced weapons was incalculable and protecting itself from industrial sabotage, not to mention competition, came a very close second.

As the crisis moved toward what everyone knew would be mutual destruction, Olivetti was entering the electronics race with what might have looked like insane heedlessness. This was not quite the case. Everyone at the Olivetti electronics laboratory knew that, when Italy surrendered to the Allies at the end of World War II, among the provisions was an agreement not to manufacture, produce, or construct weapons of war; similarly the import, export, and transit of such weapons was prohibited. However, since the company was developing computers for civilian use, those in charge at Olivetti assumed there would be no conflict. The main goal was, as it always is in business, to stay ahead of the competition. The idea of a desktop personal computer may have been nascent in the back of Roberto Olivetti's mind even then, or perhaps Mario Tchou's. Nobody had ever done it. But for the moment, what counted was to come up with a better mainframe computer; cheaper, smaller, faster, and if possible all three.

When Pier Giorgio Perotto was handed a hammer and file and told to build something, he was rather taken aback. "[B]ut I accepted it as one of the many odd things about the place," and he set to work. He was putting the final paint touches on a console when Mario Tchou happened to pass by and remarked, in his pleasant way, that perhaps Ing. Perotto could be put to better use. Perotto wrote, "And so it was that I was given the project of developing a small electronic machine that was supposed to convert [some] perforated tapes, made by a new generation of mechanical accounting machines built in Ivrea, into punch cards," which could be used in computers. He went to consult with the engineers in Ivrea. That was when he began to understand the growing rivalry between the mechanical and electronic divisions that would burst into open hostility in the next few years. The designers in Ivrea were not interested in cooperating. They considered the Pisa crowd a few idiots "chasing butterflies" and did not see why they should do anything to help—a warning to management that, even at this early stage, employees saw the new development as a threat to their livelihoods. Mario Tchou tactfully suggested that the idea of cooperating with Ivrea should be shelved and Perotto should invent something himself.

Perotto wrote, "Even if it meant I had to work alone, the idea of being wholly responsible for a project gave me a special charge." The Milan fair of 1958 was coming up and they were all working with this as a self-imposed deadline. The long-range goal was not to assemble a machine from ready-made parts but design and build their own. Roberto had given Mario an assignment with a time limit: a minimum of three years and a maximum of five, to come up with the first mainframe computer in Europe. As it happened Tchou's team, which grew rapidly from about fifty engineers to several hundred and finally a thousand, had produced its Elea 9003 by 1959. Perotto's tape card converter was an essential early step and it was ready for the fair.

Lucius Borielli, who had some engineering experience in radio and television, was among the first to be hired. He gave Mario Tchou credit for the rapid advance. Tchou, his shirtsleeves rolled up to the elbow, "was by and large the best manager I ever met." He had the rare ability to perceive someone's abilities and encourage them, inspire them too, so that they were willing to work for as long as it took. Some of the best ideas were achieved after hours. "Mario and I found

*Roberto Olivetti (left) and Mario Tchou, architects of
Olivetti's move into electronics research, in 1960–61*

it easy to be friends and we would spend our weekends together rather
than go home. Because of the element of mystery surrounding the
place, my wife told me that people in Pisa thought we were trying to
build an Italian nuclear bomb."

Another big hurdle to surmount was the vacuum tubes in common
use for the big machines, IBM included, and all the trouble they
caused. Transistors were clearly the solution but the technology was
in its infancy, with low performance and other limitations, so it had
been confined to things like portable radios and considered a novelty.
Sometime in 1957 or 1958 Tchou decided that the lab had to switch
completely to transistors. The lab had already produced two earlier
versions of the Elea using vacuum tubes with unsatisfactory results:
something had to be done. But how to improve the performance for
the largest and most ambitious version, the Elea 9003? Obviously
they had to set up a new company. They called it the SGS, and built
their own transistors. Franco Filipassi, another early pioneer in the
electronics lab, said, "It really was a world in action. Somehow we
were right at the front of this movement. We made discoveries on a
daily basis. It was the experience of a lifetime."

Ettore Sottsass, an Italian architect and designer who would also play a pivotal part in the early days of the electronics lab, had also been discovered by Giorgio Soavi. Sottsass remembered Soavi as "one of the best-looking young men I had ever seen, very elegant and racy with that faintly sleepy, almost lazy look, and the pet habit of calling himself 'Sottsass Junior.'" Given the task of finding an industrial designer for Tchou's lab, Soavi decided upon Sottsass because, he said, he was exactly the right person to invent the future. Soavi thought Sottsass would appreciate the poetic charm of Tchou, a gentle engineer who combined rare talents: the wide-ranging interests of an inventor with the accuracy of a scientist. As for Roberto, it was the beginning of a long friendship. Now in his twenties, he was developing a lifelong passion for architecture and art, and would eventually assemble an eclectic and enviable art collection of his own. At one time he commissioned an apartment in Milan that had no walls at all except for the toilet, and where various intimate accoutrements could be seen from the living room. Just by watching Tchou in action as a competent manager, he must have picked up some techniques by osmosis. But more than that: they were constantly in each other's company and would, after a few whiskies, spend endless evenings talking, not just about making and selling things but about the culture of life, of existence. The kind of discussion, Sottsass remarked, "that could only happen at Olivetti."

A documentary film on the making of the Elea 9003 by Pietro Contadini provides important insights into the experiences of those who worked there, including Sottsass. The scene is set in a laboratory. Sottsass enters, a short, balding, white-haired man with a white mustache and a discreet pigtail bisecting his shoulder blades. He wears a white open-necked shirt and comfortable gray sweater. Everything about his demeanor suggests calmness and measured contemplation as he gestures with his large and capable hands. He returns to the moment when he first discovered the enormity of the challenge.

"I said okay, but I went away terrified, because I did not know quite what to do. . . . Designing any old cabinets did not really interest me. I kept thinking about this environment" (here he referred to the attendants wearing identical white coats) "with these mysterious electronics and how to arrange the electronic mystery."

He was alluding to the ranks of cabinets, row on row, adapted from telephone exchanges, standing almost ten feet tall. "It was like living in a room full of wardrobes." His first thought was to lower the height so that the staff could see over the cabinets rather than being barricaded inside them. Raising the floor to accommodate the connecting cables was common, but Sottsass thought it would be done more cheaply by running the connections above the machines.

The central issue was the feeling that one had entered an icy, whirling void without borders or barriers, a nothingness on the precipice of limitless and terrifying space. That would be exactly wrong for the customer. But perhaps it was appropriate for the people directly involved. He decided to redesign the cabinets so that they looked even more frightening, "very severe in a sense, behind which was a mysterious divinity." So he used black metal structures which were given aluminum doors. When placed edge to edge the result would be "some sort of absolutely mysterious silver architecture; there was nothing to be seen except a block of shiny, metaphysical metal, a fairly nasty and in any case incomprehensible dynasty." Then Adriano was invited to come and judge the result. "[H]e apparently stood for a quarter of an hour in front of the closets with all the staff lined up," Sottsass recalled. Sottsass, waiting for what seemed like an eternity, "couldn't even feel himself existing any more." Finally Adriano said, "Alright Sottsass, now please be quick," and left.

Sottsass went on to have a fruitful collaboration with Olivetti for the next ten years but the relationship almost ended in 1961. On a trip to India, Sottsass became ill with symptoms Italian doctors could not identify. Barbara Radice wrote, "Ettore's blood kept losing substance and turning into water. He was living with an intravenous needle in his arm." One day his doctor came to tell him he had better make a will and, as a last-minute hope, suggested he consult Prof. John Luetscher, a famous endocrinologist at the Palo Alto University hospital in California. But the cost of treatment would be prohibitive. He was close to death. Within two days Roberto had obtained visas, passports, and tickets and opened a bank account for him with

Roberto and Anna Olivetti take Ettore Sottsass
out boating near Portofino, 1962.

unlimited credit. Sottsass was transported to California, immediately given experimental treatment, and gradually recovered. Roberto saved his life.

The Elea 9003 of 1959 was named for a new and radical school of philosophy that flourished in the small southern Italian town of the same name in the fifth century BC. Its principal exponent was Parmenides, who taught that being and existence were the true reality and that the changing world, as experienced by one's senses, was an illusion that could not be known at all. Parmenides's view of the ultimate nature of reality would form the basis of Plato's philosophy and later the foundation for the Gnostic theology of God. While other manufacturers were calling their inventions by numbers, Adriano characteristically wanted a name that summed up, not only his deeply held beliefs, but contained within it a reference to the startlingly new philosophy that Parmenides personified in the ancient world.

In less than two years his scrappy, pugnacious, free-thinking electronics division had produced, not just Italy's first mainline computer, but the first fully transistorized one in the world. Its computing power, that of approximately eight to ten thousand instructions per second, was higher than that of its competitors for several years. It was followed a few months later by IBM's own fully transistorized computer, and a year after that, by a series of IBM models called the 7070. The first Elea 9003 was installed by Marzotto, the textile company, and the second, by a bank, the Monte dei Paschi di Siena, the oldest surviving bank in the world (founded 1472) and a prominent commercial and retail institution. Some forty customers would lease this particular model. Tchou was already planning improved versions and the idea of an even more radical invention was brewing in Roberto Olivetti's fertile imagination.

There was, however, an apparent lack of financial support on a national level that was particularly galling to Olivetti, and not just because the Treasury had been presented, as a gift, an Elea 9003 of its own. The irony was not lost on Mario Tchou. He remarked in an interview with the newspaper *Paese Sera,* "Nowadays we have gained the same qualitative development as our competitors, but they receive consistent financial support from their governments. Electronic research, mainly for military applications, is generously funded by the U.S. Also Great Britain invests millions of pounds in the field. The Olivetti effort is huge, but other companies could hope for a better future [than we do], because they receive public funding."

The reproach fell on deaf ears. Meanwhile Adriano, who, in the months before the computer's debut, could have been devoting his herculean energies to arranging for public subsidies, was otherwise engaged. He had not given up on the idea of his new movement, merely postponed launching it, and had decided that the 1958 elections presented a more promising opportunity. In the first place his old pal Allen Dulles had become chief of the CIA in 1953, so now he had a friend in very high places. He consequently launched a barrage of letters and sought interviews with Dulles and others in which he attempted to make several points. He did not like, or approve of, the Communist Party because of its insistence on state capitalism, which Adriano believed was the beginning of totalitarianism. His Comunità, as he never ceased to point out, would be formed from the bottom up. Second, he intensely disliked the fact that his own party, the Socialists, had thrown in their lot with the Communists. With his usual optimism he argued that, if his party could be persuaded to separate from the Communists, this would redress the balance as the Americans hoped. His letters are on a single theme, the money to launch Comunità. He thought he might get funds from the Ford Foundation. (No luck there.) He actually approached Gianni Agnelli and received a form letter reply. Adriano confidently expected lots of influential help from Mr. Dulles of the CIA.

During Eisenhower's eight years as president (1953–1961), Dulles had the heavy responsibility of containing the USSR "without massive spending on U.S. conventional forces or risking nuclear war," as the CIA biographical entry on Dulles reports. His mandate also meant contacting social, labor, cultural, and student groups in Europe that were clearly anti-Soviet. That included supporting, or at least making encouraging noises at, Olivetti, whom he liked to refer to as "the charming Utopian Olivetti." According to Paul Ginsborg's indispensable history of postwar Italy, as the 1958 elections drew closer politics was in more than its usual flux. Olivetti wanted to split the Socialists off from the Communists and restore their independence. Amintore Fanfani, head of the leading Social Democrats, wanted to split them off as well, but in order to attach them to his own party. He proposed an "opening to the left" that would give Socialists prominence in forming a new government. That surprising idea had unexpected consequences. A strong faction in Fanfani's own party objected, and replaced him in 1959 with Aldo Moro by a narrow margin. The next person to be replaced was Moro, who lost to a fifty-nine-year-old lawyer, Fernando Tambroni, a year later. Curiously this new change of government came about on the very day, February 27, 1960, that Adriano Olivetti died.

During the election campaign Olivetti was tireless, going from one city to another, from Sardinia to Lucania, and from Rome to Turin, determined to do his utmost. Renzo Zorzi, a well-known racing car driver, who was a friend, said, "It was, for those who knew him, a heart-breaking campaign. He had no hope of success. He was shy, bashful, totally incapable of demagogery and accustomed to a language"—he probably meant that of one engineer to another—"that

was useless for an indifferent electorate." Olivetti believed, against
the evidence, that calmness, logic, and reason would be enough to
persuade his audience and had no idea how wrong he was. Instead of
the three senators and between seven and nine members he expected
(enough to act as a moderating force in Parliament), Comunità won
a single seat, his own. This was a desperately disappointing result for
a goal to which he had dedicated much of his life since his exile in
Switzerland. He might have expected, with reason, that the United
States would be sympathetic to his shining vision, and that was the
reason for the avalanche of letters to Allen Dulles and anyone else he
could think of. Some aspects of this CIA correspondence that have
been released show that Olivetti made repeated attempts, not only to
reiterate his philosophy but to include shrewd assessments of current
Italian political thought (and why his was so much better). Olivetti
probably never knew that his goals and prospects were also being
reported to the CIA by someone close to him. In short, there was a
spy in his midst. He or she "knows Olivetti well, and has studied the
Communita [*sic*] movement at first hand," is stated in a CIA interof-
fice memo dated December 5, 1957. This person spoke idiomatic
English.

Olivetti probably did not also know that, for several years, he had
been described by Dulles with the kind of amused tolerance with
which one views a dotty uncle and hopeless dreamer. Olivetti's poor
political results would have confirmed that assessment. From an
American point of view, he was perfectly useless. This seems to have
been the moment when whatever goodwill that might have existed
between Olivetti and Dulles evaporated. It is doubtful whether Olivetti
ever knew this. The last letter he wrote, on that fateful Saturday in
February, was to Dulles—and he was still making the same arguments.

આ

If Adriano Olivetti's life as a businessman and sometime politician
was in constant flux—he briefly served as mayor of Ivrea—so was
his private life. Sometime after marrying Grazia and the birth of
their daughter Laura, or Lalla, problems developed. Domenico de'

Liguori Carino, who became Lalla's husband, described it as a clash of cultures. He said, "Grazia had a problem, not just with Adriano, but all the family members. She was a girl from the petit bourgeoisie. There was a big difference between her limited upbringing and the socially and intellectually haut monde the Olivettis inhabited." She was a poor choice for the kind of wife he had hoped to get, i.e., "a homemaker who would become an Olivetti, working to further the family's interests."

The first year of the marriage of Domenico and Lalla in 1974, they were expected to spend every evening after dinner with Grazia—Adriano had died by then. It was a very formal occasion; they were expected to sit every night in the same chair, and every night Grazia would make the same joke. "Listen, that's Adriano's footsteps coming down the stairs." Grazia did not really believe Adriano was about to enter the room, but it was half in fun, half meant to provide her daughter with some crumbs of comfort.

Domenico de' Liguori Carino continued, "When Lalla was small Grazia got pregnant again, but had a miscarriage. She told her daughter it was her fault because she had had to carry her around. She also told her she always wanted a boy and that she had a name for the one who never arrived: Emmanuele." He said, "I do believe her mother destroyed Lalla psychologically."

Some three or four years after their marriage, a pretty Swiss girl was hired to be Lalla's nurse. Her name was Heidi. Franco Ferrarotti said, "Heidi was a marvelous girl, tall, quiet, smiling, not educated or cultured, but with a sweet human wisdom. She was a wholesome human being." Pretty soon Lalla fell in love with her and so did her father. Adriano and Heidi would take Lalla to Forte dei Marmi every summer and Grazia would remain in Ivrea, apparently without complaint. Her son-in-law believed that Grazia tacitly accepted the arrangement because she was attracted to another man. He also believes that Adriano knew about this liaison and thought it was an ideal solution. He even offered to pay for a trip to America together so that they could find out, as he put it, whether they were "right for each other."

Milton Gendel, who fell in love with Adriano's daughter-in-law,

Vittoria, began as a writer but moved quite quickly into the role of Surrealist photographer, and also became known for his behind-the-scenes portraits of the famous. One of them included Queen Elizabeth, in a kilt and with her hair in a scarf, feeding her pet corgis at Balmoral Castle. Gendel had served in World War II, arriving in Naples on a Fulbright scholarship in 1949. He went on to work for the Marshall Plan (sparking rumors, according to *Vanity Fair,* that he was a spy). Shortly afterward he took up photography and never left. When he met Olivetti in 1952 he was living in what he thought was "a grander, floor-through apartment" in a sixteenth-century Roman palace. Those were the days when he was penniless and keeping up appearances meant, in this case, selling off his possessions. One day when he was entertaining Bruno Zevi, the Italian architect, his guest looked around, Gendel said, "with considerable distaste because I had furnished my apartment with odds and ends and said, 'Do you live this way?' I replied, 'Yes. Last week I had to sell my automobile and in fact, we just ate the fender.'"

His financial fortunes changed dramatically once he joined Olivetti, where he was paid handsomely for doing not much. His sole responsibility was vetting Olivetti's writings in English to make them more idiomatic. One time, Olivetti asked him to look over his application for a grant from the Ford Foundation. It was six pages long. Gendel took it home and reduced it to a single page. Next day, Olivetti looked at it. He said, "'This is so admirable, it's so concise, but you haven't left me any latitude.' No, he didn't get a grant."

Despite their stylistic differences, the two men soon became good friends. All would have been well, had it not been for Gendel's new love affair. Vittoria was still, after all, Adriano's daughter-in-law so Gendel went into his office to resign. "I said, 'Ingegnere, I think I ought to leave.' He went right on talking, ignoring what I'd said. Finally I said, 'Perhaps you didn't hear what I said?' He said, 'Dottore Gendel, I have no talent for dealing with family affairs. Could we return to our business?'"

After a few years Grazia became tired of the arrangement. As her future son-in-law put it, "There were a lot of secrets that nobody talked about, but everybody knew." This was the moment when Heidi was fired. It might also have been the moment when Grazia's mother called up Franco Ferrarotti and demanded that he write articles condemning Adriano, the great lover, for seducing his daughter's baby-sitter, "which I of course refused to do." Heidi returned to Switzerland, but the affair continued. In early February 1960, she might just have discovered she was pregnant. When Adriano Olivetti left Milan that Saturday afternoon he would be making a detour to Gstaad, where Heidi was acting as a family nurse. He had been writing her letters for two years, telling her they would marry. He would make it happen somehow. As for Lalla, then nine years old, she had a particular reason to remember her father's death with anguish. She told Domenico she had quarreled with him just before he left. She thought his death was her fault.

That Saturday evening Heidi waited up for the lover who never arrived. She eventually put in a call to Grazia, full of concern. "Adriano hasn't arrived!" Franco Ferrarotti said, "Grazia roared into the phone, 'Adriano è morto!' in a fury. I tell you as a writer—it is worthy of a Greek tragedy."

12

High Stakes

On his trip to America in 1925 one of the few companies Adriano Olivetti was not allowed to see was in Hartford, Connecticut. It was the exemplary Underwood, undisputed leader in the world of typewriters and, just then, the door through which he was denied entrance. It drew him like an obsession. For a whole day he walked around the exterior of the rambling red-brick factory, pacing its perimeter as if, somehow, he could discern what might lie hidden within it, staring at the brick walls as if they "held a secret I felt impelled to discover." Now, thirty-four years later he had arrived at the pinnacle of success: his company was being offered the chance to buy the most famous typewriter company in the world.

That chance arrived at an opportune moment. The disastrous showing of Comunità in 1958 had not just been a wrenching personal disappointment but had given the critics inside his family circle an opening. "After several stormy scenes and an angry exchange of letters, the board of directors met and, though confirming Adriano as president, named Giuseppe Pero (Adriano's longtime and loyal chief financial officer) as *amministratore delegato* (managing director) with the responsibility of restraining Adriano," *Fortune* magazine wrote. "Crushed, Adriano drifted away from company affairs. Pero stepped in with an austerity program, reducing operating costs, withholding some expected bonuses and proposing some changes in Adriano's . . . system of worker benefits."

Such costs, which also involved shutting down a fund to underwrite small business start-ups in the Canavese as well as the expansive, but expensive, holiday camps, might even have had the tacit approval of Adriano himself. If the truth were known he could not bear to deprive anyone, let alone fire them. If it absolutely had to be done he would prefer not to see it happen. Meanwhile he took up his political duties in Rome without enthusiasm. There is a photograph of him in the Chamber of Deputies leaning on his desk and staring around him as if he did not know why he was there. After a few months he popped up again in Ivrea, and before long he was back in charge.

That was the moment when an executive of the Underwood Corporation paid him a visit, offering to take over the U.S. distribution of Olivetti's lucrative calculating and accounting machines. The idea was turned down fast. But when it transpired that Underwood itself was on the market, Adriano temporized. Milton Gendel said that when he asked the Ingegnere why he decided to buy it, Adriano replied, "Oh you have no idea. Underwood to me was like Mecca to the Arabs. So when I heard that it might be for sale, I did what my mother taught me to do when big issues were involved. I took the Bible and stuck a pin in it. And when I opened to the passage where the pin stuck, it said—I don't remember the exact words, but the sentiment was—nothing ventured, nothing gained. So I bought it."

The problem was that Underwood's management style had left much to be desired. Gord Hotchkiss wrote, at the end of World War II, Underwood was still using machinery that was almost fifty years old, the plant was run-down, and management had long ceased to explore new markets. Hotchkiss wrote, "successful companies . . . become complacent. They generally don't start looking for the next big thing until the existing product line (or lines) begins to falter—and by then it's too late." *Fortune* magazine wrote, "Underwood's true condition, when it was finally disclosed, caused some anxious moments. During the weeks following the acquisition the Olivetti top echelon practically commuted across the Atlantic, often in a mood of deep depression." Ugo Galassi, who was to become Underwood's new president, was in bed with a fever in a New York hotel room when Adriano dashed in to see him. "[H]e shouted, 'I know you're too sick to discuss this, but we must discuss it. We are losing money. We haven't a moment to waste.'"

Adriano in his brief role as leader and sole member of his own political party, Comunità, 1958

Even so, Underwood had a name that still carried considerable weight at home and abroad. *Fortune* observed, "It had an American distribution network that it would have taken years for Olivetti to duplicate by its own efforts. And finally, it had some well-established foreign subsidiaries, notably the profitable Adrema Werke in Germany, which were worth keeping alive." In short, buying Underwood was the key to opening up the U.S. market. It was a watershed moment, the first time an Italian company had taken over an American company. In the autumn of 1959 the U.S. government immediately slapped the new Olivetti-Underwood merger with an antitrust suit, an indication that, somewhere, the development was being viewed with alarm and that it had to be stopped. While the suit was continuing, Olivetti began to make good on the payments: three installments, two by the end of 1959 and the third in February 1960, which may have closely coincided with the date of his death. Also, seemingly by coincidence, the new Olivetti Underwood company was about to make its debut on the Milan bourse the first week of March 1960 and Adriano was making final arrangements for that the day he died.

In September 1960 *Fortune* was predicting a rosy future for the future health and success of the new partnership. Despite Underwood's loss in 1959 of $14 million and the considerable sums needed to modernize the company, the fact was that Olivetti's U.S. gross of $18 million a year, combined with Underwood's gross, for the year 1959, of $75 million, would add up to a predicted total gross of $90 million. That would bring it into the same league as other American typewriter companies such as Royal McBee and Smith Corona Marchant, if not such giants in the office machine field as IBM, which grossed $1.3

What the Olivetti assembly line looked like in 1958

billion that year. If they had been critical, family members changed their minds; the mood could not have been more optimistic. In Ivrea and Milan the saying went, "Adriano has put enough meat on the fire for us to feed on for ten years."

With Adriano as its leader Olivetti was never going to fall victim to the complacency that had allowed Underwood to slide. Ever sensitive to market rumbling, even the faintest stirrings of restlessness that less acute observers did not hear, Adriano was pouring his energies into new ideas and particularly the future of the electronics laboratory. This was despite the fact that its sales so far were mostly rental fees and its contribution to profits was minuscule. In the spring of 1958, Adriano moved the lab to larger premises in Borgolombardo, a suburb outside Milan (it would move one more time). With bigger facilities, and particularly with the news that Hartford would be used to assemble computers made in Italy, the mood was euphoric. As early as 1959 Adriano was in search of new markets behind the Iron Curtain, Russia included, where the potential was enormous. He was even contemplating selling his technology to Mao Zedong's Communist China. So states an authoritative study of Mario Tchou and the Elea 9003 by Giuseppe Rao, a diplomat attached to the Italian embassy in Beijing and acknowledged expert on the subject. That was the most audacious and dangerous ambition of all.

℃∕

Only a few people knew that the marriage of the charming and seemingly unflappable Mario Tchou had run into some serious problems. He and his wife parted, apparently by mutual consent, after a few

Mario Tchou and his wife, the artist
Elisa Montessori, in happier days

The noted photographer Milton Gendel with his twins, Sebastiano, top left, and Natalia, 1959

years and he took up a new relationship with a young artist, Elisa Montessori, whom he promptly married; two daughters soon arrived. As for Roberto, his personal life was similarly in flux but, to date, there had not been any resolution. Gendel takes up the story here. For him things were going along beautifully at the office but less well at home. Vittoria, who used to complain to him about what was wrong with Roberto, now began complaining to their friends about what was wrong with him, he said. In the middle of it all she discovered she was pregnant, and would give birth to twins. At first she said he was the father. But then she decided to call Roberto the father. Since their relationship had ended some time before, this required a certain subterfuge on her part. So Vittoria, in St. Moritz, summoned Roberto, with no doubt a teary explanation. They spent a weekend together. Sometime after that, she called Roberto to give him the news.

Gendel continued, "She told me, 'This child has nothing to do with you,' after all. I learned later that her parents summoned Roberto and Vittoria to a meeting. Ricardo, Vittoria's brother, was there as well." At that point Vittoria would not say who the father was.

"So Roberto took Ricardo into another room and said quietly, 'Chi è padre?' and Ricardo said, readily enough, 'Oh, it's that Gendel.' Roberto is supposed to have said, 'Could be worse.'" That seemed to settle the matter. Roberto had fallen in love with Anna Nogara, a brilliant young actress, and they would marry in 1960. Some time in 1958 after the twins made their appearance, Roberto and Milton met by chance at a party in Rome. Gendel said, "We were loading our plates and Roberto said, 'Let's go into that corner and talk.' So we sat down and ate. Then he looked up and said, 'Now let's talk about *our* children.'" Gendel smiled. "I thought that was rather grand."

After Adriano died and Giuseppe Pero had replaced him, nothing much changed at the electronics laboratory. Pier Giorgio Perotto wrote, "Roberto Olivetti was the only representative from top management who followed our work with attention. [A]fter his father's death he . . . helped to protect our activity from the insistent voices and controversy that questioned the validity of our existence." Roberto Olivetti was keenly interested in every development "and this created an atmosphere of friendship and solidarity between us, almost as if we were a group of conspirators." Roberto and Mario continued their evening conversations, usually over dinner at Mario's house, which had been designed and built by Sottsass, his first venture into architecture.

Elisa Montessori, who is distantly related to Maria, the renowned educator, followed the conversations with the same lively interest. She was particularly knowledgeable about contemporary politics since her grandfather, Meuccio Ruini, a senator for life, had cofounded the Partito Democratico del Livorno in 1943. This family connection naturally led to a somewhat sophisticated, not to say jaded view. She considered for instance that Adriano Olivetti thought like an industrialist, and did not understand that Italian politicians "think like Machiavelli." She also recalled that her husband was talking then about building a computer for the consumer market a decade before anyone else. She said, "Mario would show me a box of matches and say that, one day, all of this—meaning the huge Elea—will be put in a box as small as this."

Giuseppe Rao has established that Tchou, picking up where Adriano left off, had contacted the Chinese embassy because it had initiated studies of computers. Tchou was planning a trip in the summer of

*Inauguration of the Olivetti Underwood plant
in Hartford. Among the celebrants: second from
left, Roberto Olivetti, and Gianluigi Gabetti*

1961, and clearly Beijing had to be on the itinerary. When Roberto
and Anna, who had not yet had a honeymoon, learned that Mario
and Elisa were going to China they decided to join them. The plan
was to set up a rendez-vous inside China, probably Beijing. But there
was a problem for the Olivettis. They planned to make the U.S. their
next stop and already had the visas. But they were told that if they
went to China first, "the Americans wouldn't let them in." Mario and
Elisa had rather different problems. They were also advised, perhaps
by the Italian authorities, that they could enter China but would not
be allowed to leave, the argument being that Mario's highly desirable
technical expertise and Chinese parentage would repatriate him for
good. So they settled on a meeting in Hong Kong.

Roberto and Anna arrived in Hong Kong on the agreed date, but had to wait almost three weeks for Mario and Elisa to appear. Elisa Montessori said, "We thought of several improbable ways to get Mario into China without his having to show a passport." Since Roberto was an expert sailor one idea was to smuggle Mario into the country in a boat. Anna said, "We didn't have children then so we said, 'Let's go.' It seemed very funny." Elisa took a more sombre view. She told Mario, "You can go but I have our children [back home in Milan] to think of. What would become of them without me?'" So in the end Mario and Elisa did not go to China and the Olivettis did not go either. They made their separate ways back to Milan. On November 7, Roberto had not yet returned. Mario was just back via New York, so Anna asked him to be her escort for an evening at the theatre. She recalls it was a play by Bertolt Brecht. During the interval a friend joked, "Anna went to China and came back with a Chinaman," and everyone laughed. Just two days later, and by a curious coincidence, a year after Adriano died, Mario Tchou was dead. He was just thirty-seven years old.

There was a reason why trips involving China and the U.S. were particularly tense just then. The Soviet Union broke what had been a nuclear test moratorium in October 1961 by detonating a fifty-megaton bomb, the largest ever tested. It was also testing two new ICBMs and, William Perry wrote, in the U.S. "technical intelligence gathering became an even higher government priority." That same month a confrontation took place in Berlin that presented an immediate crisis. A long-simmering dispute between Russia and the U.S. over access to East Berlin had grown more and more heated. One morning in August of that year, Berliners awoke to discover that a barbed wire fence had been erected overnight to prevent movement between the two sides. This was soon fortified and became the infamous Berlin Wall, complete with floodlights, guard towers, and dogs. It was followed by a serious incident at Checkpoint Charlie. American diplomats tried to enter East Berlin and were stopped. The Soviet Union drew up tanks on their side. The Americans drew up tanks facing them just a hundred yards away on their side, guns at the ready. By then President Kennedy had tripled the draft, was asking Congress for billions of dollars in new military spending, and, most ominously, wanted funds to identify buildings across the U.S. that could be used as fallout shelters. This was the moment when, as the Office of the Historian has noted, "a wrong move during the face-off could have led to war." Neither side wanted war and the crisis was quickly resolved but the stalemate continued. The dates were October 27 and 28, 1961, barely two weeks before the death of Mario Tchou on November 9.

That morning, Elisa Montessori recalled, began like all the other mornings. Mario rose first, then she got up, took care of the children, and made breakfast for Mario. "Every day he ate the same thing, two eggs and a grapefruit." He had an appointment that afternoon in Ivrea, some two hours away by car, to discuss a major software development and improvement for the Elea 9003. But then Mario decided to go earlier because he was having problems with his printer. He was back at the house around 10:30 a.m. and left for Ivrea soon afterward. His chauffeur was a company driver, Francesco Frinzi, aged twenty-seven and experienced, as *La Stampa* subsequently reported, and the vehicle was a brand-new Buick Skylark. This luxury car had been bought for the company director and imported from Switzerland that year, one of just three in Italy at the time. It was the best-selling evolution of a convertible; a pared-down roof, usually called a coupé, had been added. Its spectacular feature was its length of sixteen feet, which was emphasized by "streamlined" styling for its sleek horizontal lines. It had all the latest advances: a plush vinyl interior—the back seat went straight across—power windows, power brakes, and a powerful V-8 engine. Its appearance on the autoroute between Milan and Turin, at a time when most Italian cars were the Fiat 500 or Topolino, which was 9 feet 9 inches long, or the slightly larger Fiat 600, at 10 feet 4 inches, would have caused a sensation. Tchou, who had work to do, climbed into the back and settled directly behind the driver.

The fastest route to Ivrea from Milan, then as now, is the Milan–Turin autoroute; one turns off at Santhià. It was a three-lane highway in those days and the common passing lane was called a suicide alley because of the frequent and deadly head-on collisions; the turn-off at Santhià was the scene of particularly bad accidents because it involved going over a railway bridge with little or no chance to see oncoming

The wreckage of the Buick Skylark in which Mario Tchou and
his chauffeur died, one day later: November 10, 1961

traffic. The only witness was the driver of a Leoncino, a heavy two to
three-ton truck involved in the crash, Carlo Tinesi, aged eighty-seven.
He said he was on his way from Turin to Milan and supposedly beginning the ascent over the bridge when he saw a Buick coming toward
him. It was in the center lane. It had overtaken another truck and was
returning to its own lane when it saw him coming. The driver tried
to return to his own lane and the car began to slip sideways. It was,
however, too close to his truck and a collision was inevitable. The
side of the Buick, right up to the seats, smashed against the front of
his truck, Tinesi said, and was demolished. Both men in the Skylark
died. The daily newspaper in Turin, *La Stampa,* which ran two articles
on the accident, reported, "The investigation of the magistrate and
the highway police of Vercelli will probably conclude that it was not
the fault of the truck driver, but at the same time it is impossible to
determine the exact cause."

Later that day there was a knock on Elisa Montessori's door and
she went to answer it, holding three-year-old Nicola's hand. Roberto
and Anna were standing outside, looking like statues. She said, "What

are you doing here? What has happened?" Roberto replied, "I can't tell you. It's too awful." His face was white. Then he said, "Mario is dead." Elisa must have staggered, because in the ensuing commotion Nicola peed in fright on Roberto's shoe. They had come to take her to Santhià. Elisa climbed into their car and it was then that the reaction hit her and she began to tremble violently. As usual Roberto had everything under control. Mario and Francesco had been taken to the morgue, the company had been informed, and top management had arrived. Elisa was spared the ordeal of identifying her dead husband; that had already been done by others and all the papers signed. They returned to Borgolombardo that evening and she was coming to terms with the fact that she was a widow now and on her own.

Like other technicians in the electronics laboratory, Giuseppe Calogero was overwhelmed by the news. He said, "The funeral was set up on the ground floor of the Borgolombardo factory, where we were producing our all-transistor computer. At the funeral six of us, including Ottavio Guarracino and myself, carried the coffin. I sobbed for the first time in my life and I seem to remember that Ottavio in front of me was doing the same. We had lost our mentor. The era of joyful work was over. Now we had to face the difficulties life was preparing for us. There would be many."

That evening, on his way back to Borgolombardo, Roberto stopped at a bar and ran into another member of the electronics laboratory. They chatted for a while and then Roberto made ready to leave. He turned, and said, "What are we going to do now?" If there was a reply it has not been recorded.

The world's first desktop computer, the Programma 101, was designed by an Olivetti team of four men headed by the brilliant young engineer Pier Giorgio Perotto. In the months following the death of Mario Tchou, Roberto had answered his own question. As for Perotto, he envisioned "a congenial machine to which one could delegate those tasks that men do badly, or which are mentally draining and conducive to error . . . [one] that would allow the storage of instructions and data," and that everyone would be able to use. It seemed like a mad

Pier Giorgio Perotto, left front, and the team that built the groundbreaking, world's first desktop computer, 1965. Also in the photograph, beginning top left: Gastone Garziera, Giancarlo Toppi, and seated at right, Giovanni De Sandre

idea but, as he wrote, sometimes in life "the craziest decision ends up being the most conservative one." He had gathered a few like-minded engineers around him and, with Roberto's immediate approval, began to deal with the formidable problems ahead. How do you reduce a memory the size of a tennis court, or even a keg of beer, and put it in a matchbox? How do you incorporate a printer? Given that punch cards had already been invented, how do you present the complex issue of programming into easy steps anyone can learn?

In the days before semiconductors, memories were produced by means of a magnetic ferrite core, "strange devices halfway between a necklace and a fabric," but they were too expensive. Then they thought of "the magnetostrictive line, a device in which information is kept . . . circulating around a ring of transmissive material. We had the brainwave of using steel spring wire, which proved very suitable." For the printer part he knew that another Olivetti designer, Franco Bretti, had been in the process of inventing a printer that linked up with a typewriter until the project was scrapped. They started talking and Bretti joined the group. "The printer and keyboard of the machine would end up being all his work."

Perotto continued, "This left the structural setup of the machine, with the problems of plates to support the electronic circuits . . . and an infinite number of things. Everything had to fit into very small spaces." Luckily he had attracted the attention of Edward Ecclesia, Olivetti's product manager for the electrical aspects of its products. The result became "an innovative technology for highly complex printed circuit boards." Creating a programming system was, perhaps, the most difficult aspect of all and took an enormous amount of experimentation before the team felt they had solved the problem.

They began work in 1962 and were still at the start of their efforts when a series of crises took place that almost prevented a desktop computer from ever happening. It all began with a market slowdown in 1962–63, a serious setback for the company, something that could not have been predicted when Adriano decided to buy Underwood. Writing about the event, Perotto does not believe Adriano would ever have bought the company if he had visited it first. When he did fly to Hartford after the sale, Adriano is said to have remarked that he should have listened to his engineers instead of his lawyers. As it was, since most of the machinery turned out to be obsolete, the plant was hopelessly outdated and inefficient, and the whole company had to be moved to new quarters a few years later.

The company's debt had climbed and in the meantime shares were plunging on the stock exchange. Perotto wrote, "From 11,000 lire at the beginning of 1962, they fell to little more than 2,900 lire in August 1963 and the price would continue to fall over the following months, so that by March 1964 they were worth about 1,500 lire. A veritable nosedive." In the middle of it all, the faithful, reliable Giuseppe Pero, who had steered the company's financial fortunes since 1920, whom all the family members loved and trusted, died in November 1963 at the age of seventy. Someone was needed to replace him, but who? At that point Silvia, who was back living in Ivrea after many years in South America, resumed her role of arbiter. In his account of how the P101 was invented, Pier Giorgio Perotto wrote that the various branches of the Olivetti family, which between them owned

Bruno Visentini

70 percent of the shares, had always been at odds over how it should be run. But while he was alive Adriano's "strong personality, his charisma and enviable entrepreneurial skills were able to contain . . . any conflicts and problems." But "all of them to some extent shared the same misgivings for Adriano's politics and ideas, which were considered too futuristic and dangerous."

Adriano's legacy was now in the hands of his young and inexperienced son who shared the same crazy ideas. With his death, fissures that had been papered over developed into chasms. Silvia demanded an end to the craziness. She trusted an old friend, Bruno Visentini, a senator, banker, and industrialist. Visentini would not betray them. Or so she thought. For the best of reasons she set the company on the road to disaster.

If Roberto was disqualified by inexperience, so was Silvia. No doubt she had never heard of a common maxim of the business entrepreneur, that by the time a company sees a change arriving, it is too late. Nor would she ever have understood the perfect storm of coincidences that, seemingly unrelated, were designed to bring the family company to its knees. It started with the European-wide depression in 1962 that saw a drop in Olivetti share prices at the moment when it was discovering the actual cost, in terms of needed investments, of buying Underwood in Hartford. Then something peculiar happened—banks seemed perversely unwilling to lend money. Even in a depression year, to refuse to lend money to a company with the size, scale, and importance of Olivetti seemed inexplicable. Roberto Olivetti spent most of 1963 traveling around Europe with the bright idea of forming a European consortium devoted to the new wave of electronics, with no luck. Camillo's heirs—the remaining shares were divided between managers and others in some way linked to the company—were asked to make further capital investments. Since many could not come up with the necessary cash, they pledged their rapidly devaluing shares as securities for loans. Dott. Alberto Galardi, a noted architect, who married Mimmina, explained, "the sudden fall in share prices brought down drastically the value of our shares. With unimaginable suddenness the banks that had financed the family for the capital increases," called in their loans. The market was abuzz with rumors that the company was going under. *L'Unità,* the Rome newspaper, wrote, "Today on the trading floor of stock exchanges, this rumor circulated: Who wants to buy a bargain? It's Olivetti. The idea that you will be throwing money away caused the company stock to fall sharply. At the end of the day the Olivetti stock had recovered six points, but very little against a fall of 238 points over two days." The newspaper reported that "the foreign press, above all in the U.S. and U.K. had been making an alarming prediction for the last week."

By May of 1964 the family had been forced out, a so-called rescue group, headed by Fiat, had taken control as the price to be paid for additional credit, and the company had agreed to turn over its electronics division to General Electric. The *Financial Times* reported that the company which had, for so long, "run what has often seemed more of a social and artistic experiment than a business" had been acquired by a hard-headed rescue group that bought it at bargain prices. Nowadays it would be termed a hostile takeover. It made for sad reading, the *Financial Times* wrote. Now Olivetti was "just a business."

The merger, or collaboration, with General Electric also turned out to be a misnomer. Perotto wrote, "At first, within the division we thought that, rather than a sale, it was some form of joint agreement. We were encouraged in this belief by a brochure from the company containing an energetic denial that the company planned to close down its electronics division." Roberto himself, "whose good faith we had no reason to doubt," welcomed the arrangement which, he thought, would give the division a new lease on life. Perotto made a different discovery when he and a delegation of colleagues were invited to visit GE's impressive research laboratories in Phoenix, Arizona, where its own mainframe computers in the GE 600 line were being developed. As he went through the laboratories, he was "seized by the most dire sense of foreboding about the fate of the electronics division in general and the activities of my group in particular." The GE engineers seemed uninterested in what was happening in Ivrea: "they did not care a fig for these small machines"; in fact, their tour guides seemed bent on demonstrating how backward the Italians were.

Perotto decided that GE was only interested in establishing a commercial base in Italy for the distribution of its own computers. This became quite clear a year later when the terms of the sale were made public. GE had bought a 75 percent interest in the Olivetti lab. After a few years the Olivetti name was dropped from the joint masthead. It was over, but Perotto and his group had long since understood.

The same was true for the takeover group. Not only did electronics have no future at Olivetti but the very idea was a blemish, a "mole" on its corporate face. It must be eradicated, Vittorio Valletta, the "Professore" at Fiat, said when the details of the takeover became public. The intervention had to be made to save the company from ruin. Perotto was still angry about the deception and dissimulation that had destroyed a first-rate electronics division. "They lied to us," he said years later.

Then the GE accountants moved in and started making a scrupulous inventory, item by item, of what the company had bought. Something had to be done to save the P101 or it was doomed. The story goes that someone, sometimes described as Roberto Olivetti, sometimes as Gastone Garziera, had the inspired idea to call it a mechanical calculator rather than an electronic one so that it would stay with the company. The ruse worked. The Perotto group remained with Olivetti instead of being sold to GE and the future of the P101 was safe. More or less. Perotto said they went overnight from being a part of a company to "visitors from a foreign land, treated as undesirables precisely by the very people with whom we should have been working closely." The future was unmistakable; mechanical engineering was on its way out and the fear and resentment on the factory floor was palpable. Perhaps industrial sabotage had already begun. The Perotto group retreated to an inconspicuous corner of the site and painted all its office windows black.

Perotto wrote, "I do not want to chronicle here all those months of work, nor call to mind the difficulties, frustrations and moments of despair . . . The important thing is that by the autumn of 1964 we had the clear feeling that we had done it. But we had never seen the whole machine work, only each part separately." So in November 1964, he loaded all the parts—now only slightly bigger than a shoebox—into the back of his car and took them to Ivrea. The machine was soon assembled, and after a few days of fine tuning began to work. The time had come to prepare a few programs and invite visitors. One of the first to be given a demonstration was Natale Capellaro, the renowned designer of the Divisumma. "We used as an example some calculations which were most frequently done manually with the Divisumma 24 and which our machine did automatically, printing the long sequence of results with the utmost speed.

"Capellaro carefully observed each process, delicately stroked the machine, as if he wanted to feel the parts pulsating beneath his sensitive fingers, and remained silent for a long time . . . [Then] he patted my shoulder and said, 'My dear Perotto, seeing this machine work, I realize the mechanical era is over.'"

The final hurdle was to design the cover of the new machine, which, when it first appeared, looked more like the office calculator and would not have a display window attached for several years. It used magnetic punch cards, the equivalent of floppy disks, instructions that, when fed into the machine would become part of its memory and that could be printed out. Sophistication and elaboration would arrive in due course. They were off and running with something easy to use, could sit on a desk, be picked up and carried, and would, as its inventors stated, come up with results incredibly fast. It took some time to find the right designer but he finally arrived in the person of Mario Bellini, a member of the Sottsass scene, who would go on to create many such advanced designs for Olivetti and become known for his multifaceted talents. Bellini intuitively understood the impulse to create a machine that would take into account the habits, usage, and emotions of its future owners. He used a die-cast aluminum casing to minimize weight, molded tightly around its internal workings to also minimize size. His keys were elastically stretched surfaces to make ideal finger contact. The P101's "gills" acted as ventilation; a protruding "tongue" acted as a keyboard support for the hand, and the "eyelid" housed indicator lights. As a new addition to office furniture it was as sleek and elegant as it was easy to use. It would make its way into the collection of the Museum of Modern Art in New York as one of the high points of industrial design.

In the meantime news about the new machine had spread inside Olivetti, even though they had tried to keep it under wraps, if only because the design had not yet been patented. They had not tried hard enough. Ing. Roberto Battegazzorre, whose father, Giuseppe, was chief of police in Ivrea for forty years, recalled that his father, who was very reserved and usually did not tell his family about his investigations, went off one morning in late 1964 or early spring of 1965 and was gone for the day.

When he returned, he told his wife and son that he had made his first trip by helicopter, something of a novelty in those days. "Gradually the reason for that journey came to light," Ing. Battegazzorre wrote.

"My father told us that the prototype of the Programma 101 had been stolen from the laboratory of Olivetti a few days before. He managed to find out more through an informant and discovered that the computer was on its way to Switzerland." In those days there was a lively traffic in stolen goods between Italy and Switzerland. Groups of smugglers would carry the merchandise, often cigarettes, over the Alps using well-worn paths at high altitude. The trip might take several days but was usually worth the trouble once the goods arrived at their destination. In this case the problem was that several days had already gone by, the smugglers could have already arrived and it would be much harder, if not impossible, to trace the prototype after that. There was no time to waste. A helicopter was hired, the police chief and his informant got on board, and so did the head of security at Olivetti. He carried a briefcase of several million lire and, no doubt, a gun.

They were in luck. They spotted the band on the trail, made the arrest, handed over the money, recovered the precious prototype, and no one, including the press, was any the wiser. What this incident revealed was that someone outside the company, whether a business

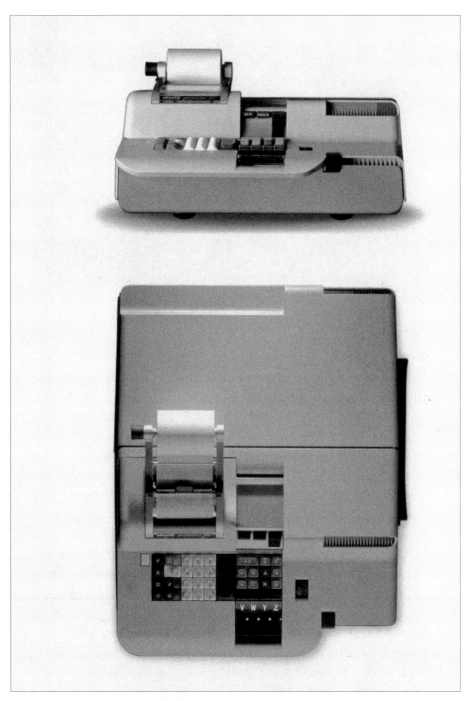

The P101, the world's first desktop computer: top and side views, 1965

*Giuseppe Battegazzorre, at the policemen's Christmas
party in Ivrea, 1958–59, with Adriano Olivetti*

rival or a foreign government, very much wanted to know what was
under the cover of this new machine. And somebody within the
company was spying on every move Perotto and his little band were
making. P101's creators had endured some anxious moments. They
had almost lost their precious prototype. Again. Now they had to keep
it constantly under guard if it was ever to be put on public view. Their
goal was to show it at the New York World's Fair, which had its first
six-month season in the summer of 1964, and was set for its second
in the spring and summer of 1965.

Visentini and his executives were making plans for that too, but
they did not have computers in mind. At the moment when the
electronics division was being scrapped the company was investing
its energies in the design and production of two mechanical office
machines. The first was based on a mechanical printing calculator
called the Logos 27, and the second, an accounting machine based
on an extremely ingenious, purely mechanical programming system.
Perotto wrote, "In a way these machines had borrowed some concepts
of organizational logic from electronic calculators, but translated
them in terms of metal: rotating shafts, ratchets and gears." The new

machines were extremely complex and needed exquisitely fine tuning to work at all. They also required an enormous investment in machinery and buildings. Natale Capellaro was not directly involved in the design but he knew from the start that it was quixotic, to say the least, because there was no way that shafts, ratchets, and gears could possibly compete in terms of speed with electronics. The whole idea was going to be a very expensive failure. Nobody was listening.

The New York World's Fair was always described in superlatives: 80 nations took part, there were 140 pavilions, 110 restaurants, 45 corporations, and it covered almost a square mile of Flushing Meadows Park in Queens, New York. American corporations and their products naturally took the spotlight. The themes, "Peace Through Understanding" and "Man's Achievement on a Shrinking Globe in an Expanding Universe," were carried out in terms of almost comic-strip Futurism, set in a landscape of parks, walks, lakes, and fountains, dominated by a stainless steel model of the earth twelve stories high. Over 50 million people visited the fair during its two years of existence in the summers of 1964 and 1965, to marvel at the spheroid-shaped pavilion (IBM), its fiberglass tower (7-Up), and its Saarinen-designed restaurants with their innovative fiberglass domes and furnishings.

Olivetti's stand consisted of a huge display of Logos 27 mechanical calculators, adding machines, accounting machines, invoicing machines, and typewriters. In the summer of 1965, the P101 made it to the fair but only just. It was hidden away in a small room behind the main stand. If anyone found it, it was supposed to be an accident. But a few people did. Soon more and more began to squeeze their way in. They wanted to touch the machine. They wanted to feel it. They thought it must be connected to a big machine hidden behind a wall and were amazed to learn it was not. They wanted to play with it. Pretty soon Olivetti salesmen who had been hired to demonstrate the mechanical machines were diverted to the P101. Extra attendants had to be brought in to manage the crowds. Perotto himself, who was always hovering in the background, was pressed into service to show the computer could play games as well. He invented one on the spot and to his delight, he lost. Perhaps the ultimate compliment that autumn was a request from NBC for five P101 computers so as to calculate election results that would be transmitted to millions of viewers in the greater New York area.

Newspapers and magazines such as *Fortune, The New York Times, The Wall Street Journal, BusinessWeek,* and others groped for ways to describe the new machine. "The first desk-top computer in the world," one said. Another, "the machine is filling the gap between the large conventional computer and the desk calculators." A third, with some prescience, predicted there would be "a computer in every office even before there are two cars in every garage." A brilliant idea that evolved in the minds of Roberto Olivetti, Mario Tchou, and Adriano Olivetti a decade earlier had led to this: universal acclaim and so many phone calls and letters that the company was overwhelmed.

The next problem, even if a gratifying one, kept the team up at night. They had assembled about a dozen demonstration machines but going into mass production was another matter, and the problem of logistics was urgent. They were concerned for a number of reasons.

The 1964–65 New York World's Fair

Their enemies inside the company would pounce if it turned out that the new machines, once tested in somebody's office, did not work, or only worked once in a while. However, the manpower to make sure they were ready to be sold did not yet exist.

This led to another eleventh-hour save that, once again, put them in the position of acting like conspirators in their own company. They waited until evening, when the plant was closed. Then they talked their way past the guards and into an area of the building where the machines, already crated, were ready to be shipped to the U.S. One by one they removed the machines and tested them to make sure they worked. "We went all through that night and the next, testing, repairing and substituting." One week later the first batch left for America "in perfect working order."

The Programma 101 was the world's first commercial, programmable desktop computer. This ground-breaking version used a keyboard, a printer unit that printed data and instructions, and also a magnetic card reader. Once programmed, this card would store data and programs. These cards were made of plastic, with a magnetic coating on one side and a writing area on the other. They were easy and economical to use; to make use of the program stored on it you simply inserted it into the machine. Its memory of 240 bytes is negligible by today's standards but was then a huge advance over mechanical calculators in terms of speed. And in terms of mainframe computers it had the advantage of being in the office and did not need a cadre of technicians in white coats. You could put it on an office desk (it weighed about sixty pounds), it was easy to use, and it was all yours. The price, in 1965, was $3,000 ($24,500 in 2016 dollars). Not cheap, but attractive enough to be snapped up by eager buyers. Some 44,000 P101 machines were sold. In short, it was a winner.

Among the buyers was the National Air and Space Administration (NASA), which was preparing to land on the moon in 1970 and bought ten copies. David W. Whittle, a programmer for NASA at the Johnson Space Center said, "By Apollo 11 we had a desktop computer . . . an Olivetti Programma 101. It was a kind of supercalculator. It could add, subtract, multiply and divide, [and] also remember a sequence of these things, [recorded] on a magnetic card . . . So you could write a programming sequence and load it in there." He continued, "The Lunar Module high-gain antenna was not very smart; it didn't know where Earth was . . . We would have to run four separate programs on this Programma 101." The computer was also being used by the U.S. Air Force during the Vietnam War to "compute coordinates for ground-directed bombings of targets for the B-52 Stratofortresses." The military had taken note.

Almost the first thing was to take out a copyright for the P101. This turned out to be a wise precaution when, in 1967, barely two years later, the American company Hewlett-Packard brought out its own desktop computer that resembled, in all essential respects, the design of the P101. They put the machinery into its own case, with HP on the label and sold it with its own number. Olivetti launched a copyright suit and was rewarded with almost $1 million in royalties.

For Perotto the P101 was just the "first lonely brick" laid toward what would be a shining road to the future. All that daring, forethought, and subterfuge had led to the startling lead that the P101 then held over every other company. Perotto thought Olivetti had a five-year advantage over the competition. David Olivetti calculated longer—seven years, a lifetime in the cutthroat business world—in which to establish a world market, making use of the new semiconductor technology that was the next bold step. It would be a world leader and Ivrea would become its center, for an Italian Silicon Valley.

Roberto was already looking ahead. He had signed an agreement with Fairchild, a company already working in the semiconductor field, to jointly sponsor further research and development, which, in short order, would lead to the development of personal computers on a vast scale. Richard Hodgson, an executive with the Fairchild Semiconductor Company, was enthusiastic. His company would gain the edge in Europe and Olivetti would gain innovative technology that would enhance its stunning advantage. All this took place in September 1959, a few months before Adriano's death. The two men liked each other immediately. Hodgson thought the young Italian was sensitive, and "brimming with youthful vitality," and Roberto saw Hodgson as Mario Tchou's logical successor. But then the objections started and the roadblocks were built by directors and sharehold-ers who, Hodgson concluded, cared only about immediate profits. Visentini scowled. Nothing came of it.

The curious, one might almost say perverse, refusal of Olivetti to seize the advantage was immediately recognized by Perotto—who would soon leave and set up a company of his own. Instead Olivetti fixated on the future of its Logos 27, a mechanical calculator that had, as was noted, to be at peak performance to work at all, and in any case was never going to compete with electronic machines. "Olivetti reacted to the first signs of the impending revolution in microelectronics with a policy of . . . passive resistance, doing the minimum possible and letting itself be dragged along by events," Perotto wrote. The company was already losing its dominant position in calculators and the American Fairchild Semiconductor, seeing the way the wind was blowing, pulled out in 1968. The company, Perotto wrote, was going down "a dead-end track." Had Adriano lived, Perotto believed "he would have promoted integration and amalgamation [of the two warring aspects of the company] and rendered it impossible to cause the reckless alienation created by his short-sighted successors." Meanwhile Olivetti continued to squander its enormous potential. The momentum had been lost.

The Programma 101 has not been well served by computer histo-rians on or off the internet. One writer dismisses the machine as a calculator (by the company's own admission), demonstrating a lack of awareness of the reasons for the deliberate misnomer. Other authors describe it as "not mass market," as if sales of 44,000 units did not qualify it for such a label. Columbia University's Watson Laboratory did build a personal computer at an early stage (1948–54) but in those days bulky systems had not been miniaturized and the one-man unit housed a massive standing box on the floor; then again, the stiff price for its machine of $55,000 put it out of reach of most. The Computer History Museum considers that the prize for the first desktop com-puter should go to the British Kenbak-1. But this machine arrived

*Roberto Olivetti, right, at a dinner
with Richard Hodgson, a pioneer in
the field of semiconductors, 1959*

six years after the P101, in 1971, and only sold about 250 units. As for *Computing: A Concise History* by Paul E. Ceruzzi, curator at the National Air and Space Museum, the names P101 and Olivetti are not in the index. The arrival of Apple and Steve Jobs is dated 1976—it appears that Olivetti's electronics engineers had even more time than they dreamed possible to produce new and advanced models of their ground-breaking invention. As for IBM, Vice President Bernie Meyerson claimed in 2011 that his company had invented the personal computer, the IBM 610, which did not appear until 1981.

Missing from any of the chronicles is the unsung role of Roberto Olivetti. This may have had something to do with his personality. He hated giving speeches, tended to be self-effacing and preferred to do "good by stealth," as his friends said. Defending himself against unfair accusations did not come easily—he was more likely to withdraw silently. And in fact, members of the Olivetti family, now disinherited, turned their frustrations and resentments on Roberto. From their limited perspective everything seemed to point to incompetence on the part of Roberto and, in particular, his ruinous determination to push through a financially reckless boondoggle. If only he had been another Adriano! But he was not; not charismatic, not persuasive, not business-savvy. He had let them down. It was all Roberto's fault, they said. When in fact Roberto's only mistake was seeing what the company had to do if it was to survive.

The Curious Case of the Second Death

"I know. I know." Poirot nodded his head sympathetically. "You would like to clear this up. You would like to know definitely exactly what occurred and how it occurred. And you, Dr. Gerard? You have said that there is nothing to be done—that the evidence is bound to be inconclusive? That is probably true. But are you satisfied that the matter should rest so?"

—Agatha Christie, *Appointment with Death*

There was no shortage of reasons given for the failure of Olivetti to capitalize on its success at this moment of triumph. Perotto thought it had to do with the shortsighted attitude of management, which clung to "an archaic, old-fashioned vision of power." For others the cause had its origins in the autumn of 1959 when Adriano Olivetti, against all advice, bought Underwood and, as it turned out, left his heirs with a financial deficit they neither wanted nor knew how to deal with.

The ancillary argument is that Adriano, aged fifty-eight, probably died from sheer happiness after achieving a goal of a lifetime. This somewhat speculative and unlikely explanation is usually followed by the observation that he was in such poor health he would have died anyway.

The third argument seeks to explain the death a year later of Mario

Tchou. It was a rainy and misty November day and the road was slick. His chauffeur was not equipped to deal with a large American car. The brakes locked and that was it. He should not have been trying to pass a truck going down a hill.

The fourth argument: The family did not understand the future in computers, but then, nobody did. Adriano and Roberto were too far ahead of their times.

The fifth argument: Roberto was not Adriano. He did not have his father's charisma or, just as important, his family's trust. He did not put his ideas across successfully to his family and shareholders. He was weak.

The sixth: The company was about to go bankrupt. Banks naturally flee from a sinking ship.

The seventh: The problem of a large family who all have equal rights. The Olivettis should have appointed one person and let him decide for them all.

The eighth: Only severe economies and slashing of employee perks could save the company. But if Olivetti could never fire anybody, their hands were tied.

The ninth: The electronics division was an expensive frivolity that a company Olivetti's size could not afford.

The final argument: It was too bad about mechanical typewriters, but the company had had a good run. So goes the world.

Such reasoning has been used many times to explain the perfect storm of calamities that afflicted Olivetti, one after the other, in the period leading up to and following the death of Adriano. However, in the years since, enough doubts have risen to merit investigation, since none of them covers all of the facts. Many can be shown to be misconceptions and calculated misrepresentations. Others have something in common with the parable of the blind man feeling the elephant. Depending on where he is standing, someone feeling the trunk will decide the being is a snake. Or the ear: this creature is an amazingly large fan. And so it goes; the elephant is a rope (the tail), a tree trunk (the leg), or a spear (the tusk). The tale is ancient and can be found in Buddhist, Hindu, and also Jain texts, to illustrate the truism that one can be partly right and wholly wrong.

An equally interesting analogy could be drawn from *Rashomon,* that masterly Japanese film of 1950 that is without equal for its skill in depicting plausible narratives of the same event. A murder is committed; its four participants reach opposite conclusions. Somewhat in the same way, the story of the rise and fall of Olivetti has been described through the prism of conspiracy, jealousy, envy, and revenge, and especially a desire not to know, rather like a character from one of Alice Munro's short stories who "carried not noticing to an extreme."

Or, as Barzini writes in *The Italians,* a man can only try to defend himself from the tides of history, "keep his mouth shut and mind his own business." It is a curious fact that such emotional defenses have a way of perpetuating themselves down through the generations and becoming less safe, even illusory. Perhaps no one was better able to reflect the distortions set up in a game of such mirrors than the playwright Luigi Pirandello, two of whose titles are *Tonight We Improvise* and *Right You Are, if You Think You Are."* Pirandello is "forever gnawing away at the boundaries between reality and fiction, madness and sanity, past and present," John Hooper wrote. "The audience at a Pirandello play is repeatedly disconcerted and misled. Apparent certainties are undermined. Ostensible facts prove illusory. His works are, in short, very much like the experience of living in Italy."

Take the early death of Adriano Olivetti, aged fifty-eight, for instance. Nothing is more reasonable on the surface than the explanation of why he died. He often collapsed, as his father had done, from sheer mental exhaustion and the need for oblivion after months of the punishing pace he demanded of himself. He had been ill in 1950 coinciding with Lalla's birth, and it took him months to recover. He could have been taking pills for high blood pressure. Given all that was happening that day in February 1960, it is perfectly possible that he succumbed to a heart attack or a cerebral hemorrhage. This verdict by a Swiss doctor is always cited. Only one of the many newspaper accounts adds that this doctor could not be sure and recommended that the family order an autopsy. Most accounts ignore the uncertainty and state this verdict as authoritative.

That is to say, it was not questioned until recently. A two-part, fictionalized biography of Adriano Olivetti's life, *La forza di un sogno*

(The Force of a Dream), was seen on Italian television in 2013. (It has not been shown with English subtitles.) It was produced by RAI, a major network, and the leading role was played by Luca Zingaretti, a brilliant actor best known in the U.S. for playing the leading role of a detective in *Montalbano,* a lengthy series set in Sicily. The film records Olivetti's train journey to Switzerland and suggests he was murdered on the train. We find him seated in a carriage, in one of the European trains in which passengers sat four or five abreast in compartments with sliding doors. A long, narrow corridor, the width of a man's shoulders, connected them to each other and adjoining carriages. As the train goes under the Alps and through a series of tunnels, someone enters his carriage in the semi-darkness. One sees Adriano draw back, his neck exposed. The stranger does something, then disappears. Adriano tries to get up, staggers forward, and dies. In an earlier version the attack was clearly shown. It is only suggested here, in response to objections from Lalla. One does not know why but one can guess easily enough that she had no wish to invite reprisals against the family, even half a century later.

Lalla remembered, as a little girl, hearing heavy footsteps below her window each night as she went to sleep. They were the pacings of her family's permanent guard, who had an apartment over the garage. Adriano would never have made much of this, or paid any attention to the death threats over the phone that he was said to have received in the weeks and months before he died. He would have regarded the purchase of Underwood as a normal business venture and dismissed the potential dangers. But if the word was that the new plant would be used to assemble computer parts made in Italy the alarm would have been heard sooner or later in a factory full of mechanical engineers. We now know that someone in the company had been complicit in the theft of the P101 prototype and the assumption is that a willing buyer paid plenty for it. We know that Luigi Perotti, Adriano's chauffeur, was armed with a gun that he kept in a locked box in the car and that Adriano did not know about it. We know that the offices of both Adriano and Dino were rifled during the funeral. We do not know what they were looking for or who they were. No one wanted to be too upset. But someone was clearly horrified. Michele Soavi,

Adriano's nephew, who produced the documentary, was filming on location in Ivrea one day when an elderly man came up to him. "I was Adriano's guard," he said, evidently referring to the man who paced in the garden at night. "I know that he was murdered." Then he walked away before the astonished filmmaker could get his name.

❧

Murder on a train is a subcategory of detective fiction that has been used almost since the invention of trains, presumably because of its many opportunities for a quick attack and an easy getaway. *The 39 Steps,* a 1935 thriller by Alfred Hitchcock, takes place partly on a train where a man wrongly accused (Robert Donat) is fleeing for his life. There are military secrets involved and narrow escapes at close quarters. Something similar is happening in Agatha Christie's *The 4.50 from Paddington* and of course her often revived *Murder on the Orient Express,* is a model of the genre. Much of the action takes place in the corridors where opportunities for close and unwelcome contact are unavoidable. A jab from a poisoned umbrella, a dagger under the ribs, a quick and violent throttling—the opportunities for the crime writer are tempting. The assailant usually strolls away and quietly gets off at the next station, disappearing in the crowd. Curiously, the CIA created a particularly handy weapon for corridor use: the poison gun, one that mimicked a heart attack. It was shown during testimony by William E. Colby, director of the CIA, at the Senate Intelligence Committee hearings chaired by Senator Frank Church of Idaho in 1975. The gun fired a small dart containing a lethal missile, odorless and tasteless, that worked almost at once and melted into the body so that it could not be detected in an autopsy. The Russians had invented a similar weapon and are known to have used it, or rather suspected of having used it, since suspicious heart attacks of certain people raised questions only later. The story is engagingly told by Serhii Plokhy in *The Man with the Poison Gun.* Colby did not think the CIA had ever used its own poison gun but since records were so sparse and often destroyed he could not be sure they had never been used.

Speculations of this sort might have been put to rest if his widow had permitted an autopsy for Adriano Olivetti. She would not, saying that Adriano's great fear was that he would be buried alive. For that reason he wanted an open coffin for two days after his death. That wish was honored. He died on Saturday night and was not buried until the following Wednesday. In the interim, might an autopsy have taken place? The question will never be answered, along with the possibility that nobody wanted to press the issue. It is enough to know that, fairly soon afterward, Grazia took Lalla and herself on an ocean liner trip around the world and was gone for a year. Had she been advised to leave? She never lived in Ivrea again.

In *The Annals of Unsolved Crime,* Edward Jay Epstein, author of many other investigations including a best-selling examination of the Warren Commission's report on the death of President Kennedy, has identified three categories of unsolved crime. The first is a death from apparently natural causes. He uses as an example the death of Mahmoud al-Mabhouh, commander of Hamas, in 2010. There were no obvious signs of violence on the body, and the coroner in Dubai concluded he had died from natural causes. After it was revealed that he had been watched by a suspicious group before his death, Hamas insisted on an autopsy. Further investigation revealed that a muscle-paralyzing drug had been administered and the victim had been smothered with his own pillow. The coroner concluded it was a disguised homicide, "meant to look like death from natural causes during sleep." Adriano's death certainly looked natural and has never been revisited for further investigation even though its effect on the company was calamitous. The rapid death of Pero three years later falls into the same category. Were these men victims of what is called an "omicidi eccellenti," i.e., a necessary, therefore justifiable death? We may never know.

The second of Epstein's categories is a variant of the first or, "the arranged accident." He explains how difficult it can be to ascertain the cause when a plane blows apart in midair and one of its passengers might have been the target for unknown assassins (as in the Enrico Mattei case, which will be examined later). Epstein could have said the same for car accidents in Italy, a country notorious for reckless drivers, as being an even better example of how a murder can be

*The notorious CIA dart gun, a top secret until its existence
was revealed during the Senate Intelligence Committee
hearings in 1975. Shown here are Frank Church,
chairman, at left, and John G. Tower, vice chairman.*

arranged to look like an accident. In this particular case the belief
that Mario Tchou's death was carefully arranged is often raised. Carlo
De Benedetti, who assumed the role of CEO at Olivetti in 1978 and
remained there for twenty years, stated in 2013 that he believed Tchou
had been assassinated.

It is worth reconsidering the political climate on the date of Tchou's
death, November 1961, just a year and a half after Adriano's. In the latter's case the Cold War in Europe was focused on ICBMs with nuclear
warheads being put in place throughout Europe, including Turkey
and Italy, leading to the Cuban Missile Crisis in 1962. Tchou's death,
which closely follows the Berlin Wall crisis, falls into the third category
of Epstein's unsolved murders: "crimes of state in which governments
make a solution virtually impossible. What might be considered an
'obstruction of justice' if performed by an individual may be done
by a government on grounds of national security. Documents can
be sealed or expunged, witnesses can be sequestered or intimidated,
physical evidence can be suppressed, and other measures . . . taken to
protect a state secret."

As Daniele Ganser makes clear in *NATO's Secret Armies,* Italy continued to be a puppet government whose role was to protect U.S. interests for many years after World War II. William Perry emphasized the direct connection between the control of electronic engineering and the preparation for nuclear war, if it was to arrive. The parallel fear of mass annihilation unleashed massive federal spending in support of secret defense work being conducted in Silicon Valley. Jerry Brown writes in his review of the book, "As much as anyone, Perry is aware of the ways, secret and public, that technical innovation, private profit and tax dollars, civilian gadgetry and weapons of mass destruction, satellite technology, computers and ever-expanding surveillance are interconnected."

A curious case that illustrates this point took place in 1959 and eerily mirrors what was about to happen to Olivetti and Tchou in 1960 and 1961. It concerns Dudley Buck, a brilliant young American engineer working in the electronics division of MIT and, secretly, for the National Security Agency (NSA) in a number of top secret projects, ranging from spy satellites to early supercomputers. At the time of his death Buck was working for Lockheed to design rapid guidance systems for nuclear missiles.

In the cause of inventing ever faster computer connections, Buck had designed a system he called the cryotron. It had a major disadvantage—it would only work when submerged in a bath of liquid helium—and was eventually abandoned. But in 1958 everyone was "trying to build computers out of cryotrons—IBM, RCA, Raytheon, GE and the NSA most especially." During a brief reconciliation between the U.S. and USSR, a team of Russian scientists visited the U.S. and was allowed to view Buck's breakthrough ideas in missile guidance technology. Then they left.

A month later Buck, aged thirty-two, came down with a high fever and a bad cough. He died quickly afterward. His colleague Louis Ridenour, a vice president of Lockheed, and former chief scientist for the U.S. Air Force, also died suddenly, of a brain hemorrhage. He was forty-seven years old. Both men were about to attend a meeting of the NSA's Scientific Advisory Board on the subject of the supercomputer being built secretly by the NSA.

The authors of a new book, *The Cryotron Files,* believe with some logic that the Russians orchestrated the deaths of both men to stop, or at least delay, the project. Although, like most who investigate such ruthless international power games, they were unable to prove it.

The problem is as valid today as it was during the height of the Cold War, and for the same reason. Of China's theft of intellectual property, *The Economist* recently observed that what is at issue are "the core information technologies. They are the basis for the manufacture, networking and destructive power of advanced weapons systems." A country with the most sophisticated solutions establishes "an unassailable advantage." But if its technologies fall behind, it will be "squeezed out of vital technologies by foreign rivals pumped up by state support." During the Cold War the focus was on the national interest and such life-and-death achievements in the arms race. And IBM had long benefited from the government funds needed to keep it well ahead of Russia. The company was now at work on an air defense computer system with the benefit of access to pioneering research by the Massachusetts Institute of Technology. Whether in business or in weaponry, the U.S. would always depend on getting there first; "the country's biggest fear was being outclassed." Such a dream partnership not only made IBM preeminent militarily but in the domestic computer market as well. In 1964, IBM introduced a series of computers with interchangeable software and employed over 170,000 engineers and staff. But even a giant can be uneasy about an upstart company making fast inroads in a new field and the worrying possibility that it may be leaking trade secrets that are as vital for IBM to shield as the national interest.

There was yet another issue of concern under the code name "Tempest," but it was so secret that its very existence was not known for years. It concerned cyber security. It had been known since World War II that all information processing machines emit tiny radio waves that, with the right equipment, can be picked up and played back to read the messages being delivered. This matters a lot when you are in an information war with an enemy who can pick up your signals, decode them, and use the knowledge against you. What had been thought to be a problem limited to the distance of a few yards became a bigger problem soon. The CIA learned that Bell teletype machines could read text from a quarter of a mile down the signal line. Japanese scientists also discovered this. The Russians were planting not just a scandalous forty microphones inside Moscow's U.S. embassy, but installing mesh antennae in the ceilings to pick up coded messages as well. For the moment, the U.S. knew more about the dangers than how to prevent them. This, then, was the atmosphere surrounding Olivetti's bold forays into the world of electronics. Alarms were being raised in all kinds of unanticipated ways, and Adriano Olivetti and his engineers were right in the crosshairs.

It seems logical, as *La forza di un sogno* implies, that the company's entry into electronics led to Adriano's death. In the film the idea ends there, but could equally well apply to Mario Tchou. In fact, it is the second death that arouses questions about both. The company's historic acquisition of Underwood had placed it in the unique and paradoxical position of an American, as well as an Italian, company poised to sell potentially strategic information to the enemy. When Elisa Montessori was asked whether her husband knew he might be trespassing on military secrets she said, yes, he did. From an American perspective, an American company was preparing to become the backdoor conduit to the country's Communist adversaries, selling potentially vital secrets. If Tchou's movements were being monitored, which could have been the case, the news would be that he had spent the summer of 1961 trying to get into China to talk about a deal. Could that have been allowed to continue?

The move to contact Russia and China has to be seen as a political miscalculation of major proportions on Adriano's part. If he still thought that Allen Dulles and the CIA were kindly disposed toward him and his ideas, he was deluding himself. After the abysmal showing of Comunità in the 1958 elections, Adriano Olivetti went from being a possible ally to a Socialist whose party had allied itself with the Communists. That loss of influence could well have led to a series of well-coordinated and highly sophisticated efforts to stop him and his company in its tracks. Whatever the cost.

As for exactly what happened on the autoroute the morning Mario Tchou was killed, the official report holds to the theory that the Buick was trying to pass, saw a truck coming, tried to pull back into the right lane, and was hit. The caveat is that no one but the truck driver involved actually saw this happening.

The accident is reminiscent of a similar, so-called accident involving a truck that took the life of a famous American general. This was, for many years, also judged to have been the fault of his driver. The case is the death of General George Patton at the end of World War II in December 1945. Like Adriano Olivetti and Mario Tchou, General Patton had formidable enemies. They included factions within the U.S. who wanted him dead, because "he was threatening to expose allied collusion with the Russians that cost American lives," Tim Shipman wrote. Nevertheless this version of accidental death was accepted until a former spy for the OSS, Douglas Bazata, revealed that he had killed Patton on orders from the head of the OSS, General "Wild Bill" Donovan. Donovan was quoted as saying, "We've got a terrible situation with this great patriot, he's out of control and we must save him from himself."

No one has as yet come forward to claim responsibility for the death of Tchou but there are plenty of theories. As has been noted, the Buick Skylark was a high-style, luxury car bound to get noticed in a scenario of Fiat Cinquecentos, or ancient Deux Chevaux, their canvas mudguards flapping like skirts in a high wind. Since the Skylark could be seen coming for miles on the autoroute it would be perfectly possible, as in the case of the Patton crash, for the news to be radioed ahead and the moment prepared for a heavy truck to take position. Two questions immediately come to mind about the truck driver Tinesi's account. A photograph of the scene is reproduced in Contadini's invaluable documentary about the Elea 9003, the only one known. The scene shows the road itself and it is clear the accident must have taken place on a stretch of road *before or after* the bridge, because it only has a slight incline, and not the bridge itself. The road is also clearly marked for two lanes on the Buick side; the third is on the side Tinesi said his truck was traveling. Logic would have it that the Buick did not need to be passing anything, since it had two lanes, and that calls into question the assertion that the driver lost control as he was trying to move into the right lane.

This same photograph also shows the truck and car at the moment of impact. The truck is positioned to be striking directly at right angles to the car, its front bumper lined up squarely against the Buick's damaged side. Jose Zamara, founder of Zamara's Auto Body in Rockville, Maryland, who is frequently consulted by the police in such cases, was shown the photograph of the damaged car. He knows this Buick model particularly well. While still a teenager he assembled a Skylark himself. He said it had no structural frame and was fitted together piece by piece. This fact is confirmed by Ralph Nader's Center for Auto Safety manual of 1970. The car is described as structurally lightweight and with a roof that collapsed in a roll-over, killing its passengers. There were, of course, no seat belts in those days. Both front seats were designed to tilt forward to give access to the back. But in an accident such seats had a tendency to collapse forward, potentially sending their back-seat passengers straight at the windshield.

Judging by the damage to the Buick, Zamara did not think it had been hit squarely on. Neither did he think the Buick's driver was at fault, although side-swipe collisions are notoriously difficult to assess in terms of blame. He believed that the Buick was hit on its side by a truck coming from the other direction, traveling at 45 to 55 miles an hour, and rammed with such force that it destroyed the driver's door and everything behind it. Marks clearly show that the truck pulled away from front to rear, doing considerable damage to the back of the car and lifting the whole section up in the air. The evidence does not support an out-of-control turn or any other collision. The passenger side of the car was left intact and there was no damage at all to the front bumper or hood. The worst damage has been done to the driver's side and with uncanny accuracy.

There are other reasons to suspect the version given. The police verdict was that the driver was drunk. Roberto Battegazzorre said his father, Giuseppe Battegazzorre, arrived on the scene before the Buick was towed away, and found both victims already laid out in the back seat. As he opened the door on the passenger side he was hit by an overwhelming smell of alcohol. This immediately aroused his suspicions, since neither man drank. It was known that, at that time, it was notoriously difficult to get information about accidents from the highway police responsible for that stretch of road. In addition, that particular junction was known to be faulty and there was some kind of accident there every week. Employees at Olivetti knew that if involved in a car crash the price for police forgiveness in that area was a new typewriter. Fausto Lanfranco, an engineer working in the electronics laboratory, said that "We all believed Mr. Tchou was killed, that's for sure." He was asked who, specifically. "Everybody. It was a general feeling." He added, "Maybe Adriano too."

Commenting on the death of Tchou in *Capitalismo Predatore,* Bruno Amoroso and Nick Perrone observed that once the oak, meaning Adriano, was dead, the surrounding plantings, meaning Mario Tchou and his team, continued to nourish his many projects with success. This however, "was a thorn in the side of similar European and U.S. projects. This was also a sensitive area for the military aspects of security involved and as the Cold War intensified. All these good trends ended when, on November 9, 1961, this young engineer aged 37 died in a car accident. And in case the circumstances aroused suspicions among his friends and associates the event was quickly archived as an accident, just one more of those mysteries protected by 'omission,' explicit or implicit in Italy."

The sequence of events that closed down Olivetti's computer research at the height of its success devolved from a number of unfortunate factors. As has been noted, these must have had their origins in the fact that an Italian company, now allied with a famous old American firm, was in contact with, first Russia, then China, for the exchange of technical information at the height of the Cold War. What had been implicit in the deaths of Adriano Olivetti, Mario Tchou, and his chauffeur, Francesco Frinzi, became explicit when, in 1963 the so-called Rescue or Interventionist group made up of a few Italian companies led by Fiat and Pirelli shut down the electronics division. (Olivetti revived its computer program some years later, but never made up for this crucial disadvantage.)

The takeover was designed to destroy, made easier by the fact that many family members did not understand that the mechanical machines on which their incomes depended were about to become obsolete. Anyone who saw that coming would have understood that the electronics division was being sold to an American company in order to be shut down, if not why. As it happened, the businesses like Fiat and Pirelli that were instrumental in the interventionist group were simply returning favors to the Americans, another important clue; they had benefited greatly from economic aid via the Marshall Plan, and one good turn deserves another. Then there was Valletta's speech to the shareholders in the spring of 1964 in which he stated that Olivetti's electronics division was a "mole," i.e., a blemish, that had to be eradicated. Silvia, lacking expert financial knowledge, would have had to rely on the assurances of Visentini when, as David Olivetti said, "He was part of the problem." One of the few in management who knew the real story was Mario Caglieris, a high-ranking executive in the company. He was involved in the sale to General Electric in 1964 and so was privy to insider information. Famously circumspect, Caglieris kept what he knew to himself for decades. But in a taped interview with Matteo Olivetti, David Olivetti's son, in 2000, Caglieris said that there was pressure on Cuccia from outside to close down the electronics division and it came from U.S. intelligence and IBM. These, then, were the political influences to be discovered by the Harvard Business School in its probing report. If Caglieris knew then, others above him in high positions also knew. One of them had to be Roberto Olivetti, who took the knowledge to his grave. Why, Carlo De Benedetti mused, had Visentini made the company's financial situation look "much worse than it actually was?" The clues were there, but they were missed.

What is also in doubt was whether family members understood an intricate shareholder system that, as John Hooper explained in *The Italians,* benefits big business and is "almost certainly the strangest and most convoluted business culture of any of the large Western economies." It amounted to a secret pact.

As Hooper explained it, groups of large investors, usually banks or other firms, "get together to establish control of a listed company. Rarely do they need a majority stake. A third or even a quarter of the shares will normally suffice." The advantage of such a system is to give a company the financial security it needs to plan for the future. However, if interests collide among members of the pact, "it can also inflict paralysis.

"The latter is especially true if the firm in question is linked to a network of cross-holdings—stakes held by one publicly-owned company in another. These are equally characteristic of Italian capitalism. For many years the Milan investment bank of Mediobanca and its . . . president Enrico Cuccia sat at the heart of a tangled web of such cross-holdings. The influence of the secretive Cuccia reached into the clubby world of Italian capitalism, which is summed up in an untranslatable phrase, *il salotto buono*. It moots the existence of a refined drawing room where giants of industry meet the titans of finance to stitch up deals." The intricate interrelationships then existing between Cuccia, Fiat, and the Agnellis, personal as well as financial, the enormous political influence wielded by Valletta, Fiat's CEO for many years, and under-the-table deals of the 1950s that did not come to light for decades, have been well documented by Alan Friedman in his groundbreaking study of the Italian power network. Thanks to his research, the fact of a secret shareholder syndicate was made public for the first time in 1988. Friedman wrote, "this gave the Agnelli-led clique power equal to the state banks at Mediobanca . despite its negligible shareholdings." The same can be said of the ease with which the so-called rescue group took over the family control, pushed out its stockholders and directors, and put its own men into power.

Roberto's daughter, Desire, has expressed the belief that all this could have been averted if her father had been tougher. David Olivetti said, "In our world you didn't need to be tough. Adriano knew what the business world was like, but he shielded us from it. We did not discover just what the problems were until he wasn't there any more and Roberto had to deal with them." He made a reference to the predatory nature of Italian capitalism. "We weren't educated that way. In addition, forces had built up that were all coming together, and we were competing against Fiat, not for what we sold, but the way we did things, our corporate philosophy. If you go to Fiat you will find it is surrounded by walls. We didn't have any walls."

One more factor was why Italian industrialists would want to intervene, shut down one of Olivetti's most important programs, and take control of the company. The year of President Kennedy's assuming office, 1961, was also the moment for the idea of allowing Socialists into government, an "opening to the left," or *l'apertura a sinistra,* in combination with the ruling Christian Democrats. For the first time this became a real possibility. While the arguments raged pro and con, Confindustria, a powerful employers group, was bitterly opposed, warning of "ever-increasing workers' power once the Socialists came into government." Paul Ginsborg wrote, "conservative elements controlled most of the country's newspapers. Not for the first time in Italian history, crucial sectors of the upper classes turned their faces firmly against a strategy of progress, and against more equitable cooperation with the classes below them." Adriano Olivetti had shown the world what enlightened capitalism could do, and in the process aligned himself with the cause of workers and against "capitalismo predatore." Olivetti had to be stopped; but whether the family fully understood this is again unlikely.

Such a seismic shift in the political landscape was not lost on the U.S. State Department or Allen Dulles at the CIA. David Talbot, Dulles's biographer, observed that the CIA, "which had a proprietary sensibility about Italy" dating back to its successful intervention in the 1948 elections, was working behind the scenes to block the *apertura.* James Angleton's anti-left campaign went into high gear, turning Rome, as Talbot wrote, into "a key Cold War battleground." Former ambassador Luce was writing long letters to President Kennedy arguing against any softening of the U.S. position. She wrote, "Italy's pro-West government has had one foot on the Moscow banana peel for seventeen years." If the Socialists, who were pro-Communist, gained power, "the Italian Communist Party will negotiate Italy's future with the U.S.S.R." Kennedy was sympathetic to a center-left solution, but not necessarily to those serving in the American Embassy in Rome. "Vernon Walters, the military attaché, advocated armed intervention should a center-left government be formed."

There was another, more sinister reason for the effort to stop the Socialists by Italian business interests, according to Claudio Celani, a counterterrorism expert on Italian affairs. He said that Antonio Segni, then president of the Italian Republic, was in close contact with Col. Renzo Rocca, head of the military secret service, SIFAR, which was working with Borghese, the Black Prince, and Gladio. According to Celani, "Rocca reported to Segni that the financial and economic establishment predicted a catastrophic economic crisis if the Socialists joined the government. In reality, a few large monopolies (in the hands of the same families who had supported Mussolini's regime) feared that the new government would introduce reforms to break their power in real estate, energy, finance and economic planning." After leaving SIFAR, Col. Renzo Rocca went to Turin to work for Fiat.

As for the stock market, the crash of 1962 offered yet another opportunity for the "Heads I win, tails you lose" game being played. Writing in 1988, Friedman observed that the small stock exchange in Milan was still "riddled with the kind of insider trading which would by now have led to a score of criminal indictments in Britain or the United States." It is a curious fact that the period in which members of the family were forced to sell their stock at rock-bottom prices in order to repay bank loans was followed rather rapidly, once such stocks had fallen into the hands of the victors, by a resurgence in their value.

It was back to the Harvard Business School report, which established that Olivetti as a company was structurally sound and that the Underwood operation, which was ostensibly the major financial drain on the company, "was earning a profit in March 1964, a matter of months after the new group took over." In addition, "Olivetti used only one-fifth of the line of credit which had been established for it by the intervening consortium. In terms of dollars this amounted to a little more than six millions on an asset base of over 300 million." To the enraged members of the Olivetti family it seemed a very paltry return for the price they had to pay.

Harvard's investigator also visited the mysterious Dr. Cuccia but Olivetti's man got there before him. He was Dr. Nerio Nesi, director of financial services at the youthful age of thirty-three. "I remember Roberto Olivetti sent me to ask for money from Mediobanca where, he said, I would find one of the greatest bankers in Europe. He would only give me ten minutes. So I was granted an interview, and with the passion we all had then I began to explain to him what Olivetti meant, the computers, the machines, the Canavese, the phenomenon of farm laborers who became factory workers, this great culture dominating it all. After five minutes, not even ten, he stopped me and said, 'Dear Dr. Nesi, look at me. All this does not matter one iota, nothing! What matters is the value of the shares in the stock market, goodbye.' So I went to Roberto Olivetti and said, 'Is this what bankers are like?' And he replied, 'Why do you ask? You thought they were different?'"

Many years later Dr. Nesi still could not believe that the Italian government, which, far from supporting Olivetti's research, as was being done in the U.S. but also France and elsewhere, also refused to act. He said, "No one fully understood what was going on in our Electronic Division, even inside the company. I wrote letters to the Italian government to try to make them understand what Italy was losing. The issue remains permanently in my soul, in my heart, in my brain as a big missed opportunity."

When a representative of the Harvard Business School was investigating the same sale, it was a different story. Ing. Cappon, general secretary of the state-owned credit agency, the Istituto Mobiliare Italiano (IMI), said, "We came to the Olivetti situation at the request of the Banca d'Italia because it was worried. At first we thought the company situation was bad. We made a joint study with Mediobanca. We thought the corporation needed more than it did; there were a lot of rumors. We bankers think that when a crisis occurs, something is wrong. This was not the case with Olivetti. The psychological aspects were more important than we thought at the time."

Harvard's investigator also asked Cuccia why he had created a small group of private and public companies to take charge instead of Mediobanca itself or the government holding company, IRI. He said, "The politicians wanted *IRI* . . . to take over part of Olivetti," but the group was created "for political reasons."

Logic would have it that these apparently random events are linked by an underlying motive that only appears in retrospect. Perotto learned that his bold new venture had made enemies inside the company and perhaps given rise to industrial sabotage, as the theft of the prototype just before its New York debut would indicate. The rifling of Adriano's and Dino's homes during the former's funeral, a mystery that was never solved, could rank as early sabotage as well. Nevertheless in-house antagonism is inadequate to explain everything that followed. If Olivetti's race to build a computer, along with his arrival on American soil with the purchase of Underwood, had aroused concern at the highest levels, then his death two or three months later can be seen as the first attempt to close down Olivetti as a rogue computer company for good. When this did not happen, and when it became clear that Mario Tchou had taken up the challenge with astonishing success, and was now trying to link up with the Chinese, this might be enough to seal his fate. And for a secret army created and financed by NATO, there was always a cadre of Gladio's thugs, sitting around with nothing to do. Then Perotto arrived with his revolutionary desktop computer, the first in the world. Clearly, the electronics division had to be shut down for good. It was almost as easy.

Carlo De Benedetti thought there had been American, but also Russian, Pressure on Valletta to "stop Olivetti—that's for sure," he said. "Then they forced Olivetti to sell the electronics division. Nobody wants to say it but many people think like that." De Benedetti referred to the fact that IBM, the government-aligned company working hand in glove with military intelligence, had to have been just as involved. "Computing was, at that time, a monopoly of IBM and it would be unthinkable such a monopoly could be challenged by an Italian company." Taking all the factors into consideration, "it was easy to understand why Olivetti was stopped."

❧

Russian interference appears to have begun in earnest after Carlo De Benedetti took over as CEO in 1978 and brought the company back into the personal computer market with the M20 model in 1982. Roberto Battegazzorre, the son of Ivrea's police chief, said that foreigners moving to Italy were required to surrender their car license plates

Carlo De Benedetti, right, conversing
with Gianni Agnelli, 1980

for Italian ones. At a certain moment, he said, his father could not believe how many Russians were applying for new plates in his small town. Surveillance went on for years. Computers, with their design flaw that allowed them to leak coded messages that could be picked up, if not immediately read, had to be redesigned to eliminate the weak radio waves. NATO was financing the Tempest program, whose mission was to remove the fatal flaw. It transpired that Olivetti was working on a secret NATO contract to do just that, and had come up with a solution under the acronym of MSG 720 B.

The fact that Olivetti was working on such a solution was known to only a few key people inside the company. One of them was Maria Antonietta Valente, a fifty-one-year-old secretary who joined Olivetti at sixteen and had been there ever since. She was now highly placed in the company's Eastern Economics Division. Maria Valente, married, with two children, had recently been told by the company to take early retirement. What was only known to a few friends was that she was having an affair with someone in her division. He was Roberto Mariotti, unmarried, in his thirties, a high-living womanizer who was always short of money. So was Maria Antonietta, for a different reason. She wanted a facelift and surgery to shave off some unwanted rolls of fat, and look younger for Roberto.

Fate conspired to bring about just such a possibility for both of them. Maria Valente began to get mysterious phone calls on her home phone at via Galluzia 3, a few hundred meters from Olivetti headquarters in Ivrea. It seemed that Victor Dimitrev, a Soviet official with its Ministry of Trade in Moscow, was actually a Russian spy. He wanted to buy Olivetti's designs for a certain kind of explosive fuse. Valente was quite willing, but explained she could not help him, because she knew nothing about military hardware.

Then the subject of Tempest came up. Valente knew that Olivetti's files contained top secret NATO information in twenty-five folders having to do with company research. Dimitrev named his price: $250,000 in U.S. dollars, due upon delivery. She and Roberto decided to split the proceeds. The date for the transaction was set for July 1989.

The actual files would not be affected. Valente knew that duplicate copies in microfilm were on deposit at a small Ivrean company that specialized in guarding such secrets. She would get a copy of the microfilm. She contacted the company's owner, not realizing that he was a former policeman. He informed the Italian security police, and they set up a trap for Valente. She was caught in the act of making the transfer on July 10, 1989, and arrested. But not Mariotti. He had already made his escape to Moscow, where he changed his name and married a beautiful Russian girl. Valente, now dubbed Olivetti's "Mata Hari," was sentenced to four years of house arrest. On hearing the verdict, she shrugged. "Four years is no problem," she said. "So what? My husband already knows where I am."

બ∂

The successful attack in the 1960s on the Olivetti company's future development, particularly its achievements in the field of computers, was not an isolated case. As the foreign affairs officer who served in Rome pointed out, the United States had won, and acted victoriously as countries had done for centuries, assuming virtual control of Italy's political, scientific, economic, and business interests. Or, as Ing. Gastone Garziera, that canny member of the electronics division who was part of Perotto's famous team, and now guides visitors through Olivetti's small museum in Ivrea, said quietly, "Nothing happened in Italy in those days that America did not want to have happen." These were powerful forces, rewarding, cajoling, bribing, or forcing interests out.

The result was a toxic mix so labyrinthine and impenetrable that it rivaled in intricacy and cunning the network of secret cross-holdings that for decades kept the Italian business world a safe place for its biggest stakeholders. Nobody had quite been able to see how it was done. In 1964, for instance, an American journalist, Victor K. McElheny, reported from Italy that scientists "seeking an improved climate for research" were running up against "fossilized structures" in government and universities that perpetuated the status quo. What is more, McElheny wrote, these laws "have been used to prosecute several leading scientist-administrators and scientists who fear that the consequences will seriously harm research." Politicians were the easy targets. But far in the background, pulling invisible strings, were the real puppet masters, manipulating all those endlessly reflecting mirrors.

Only recently, with the revived interest of academics free to investigate and publish their findings, has there been a willingness to reexamine cases once considered closed. One of the most prominent is that of Enrico Mattei, a civil servant whose mandate was to dismantle the Ente Nazionale Idrocarburi (ENI), a monopoly on gasoline and natural gas established by Mussolini, and return ownership of the vital resource to the Middle East cartel of companies that included Exxon (Standard Oil), British Petroleum, and Royal Dutch Shell. After taking up his post, however, Mattei decided that it was in his country's interests not to close down ENI but to strengthen it. He tried to negotiate contracts with the Middle East cartel but, finding his way blocked, turned to the Soviet Union. "He not only offered to pay them in hard currency for oil but to build a pipeline through Eastern Europe so it could be delivered to Italy," Epstein wrote.

It was 1961 and the newly elected president, John F. Kennedy, was deeply concerned. A date was set for a face-to-face meeting between Kennedy and Mattei in November 1962. The two men never met.

*Enrico Mattei, murdered
when a bomb exploded in his
plane in the fall of 1962*

In October 1962 a plane in which Mattei was traveling blew up and its three occupants died. As Epstein observed, before the days of the black box flight recorders, it was almost impossible to know why a plane crashed. Mattei's death was recorded as an accident, but rumors persisted. In 1970, Mauro De Mauro, an investigating journalist, was asked to make further investigations for a documentary, *The Mattei Affair.* He disappeared and his body was never found. Two police investigators who went in search of De Mauro's presumed kidnappers were also killed, it was said, by the Mafia. Recent exhumations of the 1962 victims of the plane crash have established that a bomb was smuggled on board the plane. Epstein wrote that he believed French Intelligence responsible. As with the death of General Patton, old sins cast long shadows.

Marco Pivato, an Italian journalist, has identified two more cases in the 1960s in which pivotal figures in postwar Italy were stopped in their tracks by the same mysterious combination of forces. In *Il Miracolo Scippato* (2011), Pivato discusses the case of Domenico Marotta, a distinguished scientist and biochemist, former director of the Istituto Superiore di Sanità (National Public Health Institute), who not only reinvigorated that stately institution but built a factory for penicillin, the first in Italy. He presumably fell afoul, not only of pharmaceutical interests in the U.K. and U.S., but his own country. A campaign of calculated attacks and innuendos in the press was followed by a trial in the early 1960s. Despite persuasive evidence that the charges had been manufactured, Marotta was sentenced to six years in prison, a sentence that was later reduced. Twenty years later (1986), the verdict was finally in: "incomprehensible political revenge."

Another likely example of American behind-the-scenes manipulation of events in Italy in the 1960s is the case of Felice Ippolito. He was a geologist, politician, engineer, and early enthusiast of nuclear energy for peaceful purposes in Italy. He had graduated in civil engineering with a special interest in geology and was appointed professor of Applied Geology at the University of Naples Federico II in 1950. His research on uranium made his an informed opinion in the promotion of nuclear reactors and he was deeply involved in the Comitato Nazionale per l'Energia Nucleare (CNEN) as well as plans to build three nuclear power plants in Italy. Then, in the summer of 1963, charges were brought of administrative irregularities. Ippolito was arrested, tried, and sentenced to a preposterous eleven years in prison. "The case is widely considered as a scapegoat to block nuclear power in favor of the nascent Italian oil industry," Wikipedia commented.

What is missing from this summary is the plausible supposition that American interests were not far behind. In four crucial areas—electronics, the oil industry, nuclear power, and Big Pharma—pivotal figures like Olivetti, Tchou, and Mattei either died under mysterious circumstances, or were imprisoned during the Cold War. One can only conclude that U.S. political and commercial interests, in close coordination with competing businesses, legal and/or political factions within Italy itself, were pulling the strings.

After the huge success of the P101, Roberto Olivetti, who had first insisted it would work, then made it happen, and then showed there was a market for a personal computer, was appointed head of a new "Systems Division." It included the plant in San Bernardo where the P101 and the accounting machines were produced, and the factory in San Lorenzo, which built the teletype machines. Meanwhile, if Roberto had ever allowed himself to hope that General Electric would advance Olivetti's computer program he was rapidly disabused. The sale, 75 percent to GE versus 25 percent to Olivetti, made the relationship very clear; the disadvantage was horrendous and the whole division would be shut down by 1970. The joke around town had already begun: GE had turned into an employment bureau for Olivetti's engineers, who either understood the situation and left, or were told there was no work.

—

One of the longtime executives few recall with pleasure was Bruno Visentini, an Italian politician, senator, minister, and lecturer who had been brought on the scene at an early stage by Adriano's sister Silvia. He remained chairman of Olivetti for almost twenty years, from 1963 until 1982. During that period he was also serving important roles in the Italian government, either as minister of finance or minister of the budget (1974–87), and was frequently in Rome. But it was understood that he left a watchdog in Ivrea to make sure, as was repeatedly stated, that Olivetti would *never, ever* make the same mistakes again.

What is therefore interesting is that Visentini had appointed Roberto Olivetti, who might be suspected of all kinds of rash initiatives, to the position of managing director in 1967. David Olivetti thought he could explain that. He said, "In about 1966–67 my father [Dino] was unhappy at Olivetti. His ideas were being ignored and he wanted to start his own company, and bring Roberto in with him. I think Visentini was anxious to prevent that, so to keep him in Ivrea he gave him the job. But it was split two ways; his co-managing director was Bruno Jarach, who was Visentini's man, to keep Roberto in line."

At this moment of triumph Roberto Olivetti appeared on the cover of *Life* magazine as the new commander-in-chief, and all seemed to be going smoothly. But in this "romanzo giallo" of the Italian business world with its secret deals, conflicts of interest, back-stabbing, and dark plots, nothing ever is. Of all the equivocal roles Roberto Olivetti found himself obliged to play, this was surely the most difficult. On the one hand he and Dino were the only family members left with any influence on the firm's fortunes, with the possible exception of the aging Silvia, who, one supposed, was still receiving reassuring pats on the hand from the secret mole she, all unwittingly, had put in power. Roberto's prominence was based on the international réclame he had received from the P101 that had caused an internal uproar and which, under no circumstances, would he be allowed to develop.

He must have known he was a figurehead, a puppet, enmeshed in a much darker conspiracy coming from a direction so mysterious and powerful that no sane man would want to investigate, much less confront it, if he valued his life. Not that this would be considered anything unusual in Italy. It was back to the seasoned advice of one of the authorities on such matters, Francesco Guicciardini, a sixteenth-century Florentine lawyer, diplomat, and historian who, in a tumultuous period, possessed the sure-footed balance of a lumberjack riding an enormous log over a rapids. The secret, Guicciardini wrote, was to be truthful with truthful people, deceptive with tricky people, and the wisdom to know the difference. No wonder Roberto was studying psychology. In a world where most men were so duplicitous, discerning the difference might be critical.

In addition, Guicciardini wrote, men should at all costs keep their own interests secret, as the only way to prosper in treacherous times. Roberto, as a seasoned sailor, would have known that the verb "navigare" was often used as a metaphor, meaning a sharp eye on the shifts of the political winds and a realistic awareness of one's limited power to ride them out. Perhaps this was the time when a certain look of caution hovers at the edges of his smile, the look of a man who is already guarding too many secrets.

Roberto was probably aware it would not be long before Visentini would look for ways to get him out, and be willing to manufacture them if necessary. David, who served on the board for a few years, confirmed that Visentini was seizing on the slightest excuse to bring Roberto down. At one meeting, when Roberto was not there, Visentini "was treating Roberto very poorly and I defended him—it was over something stupid—in a democratic way—and I didn't think much about it. A week later I got a message from his secretary to come to his office. So I went there. I sat down, and he pushed a letter at me across his desk. I don't recall what it said in detail, but it was, basically, that I had not said what I said in defense of Roberto. He wanted me to sign it so that my comments would not appear in the minutes. I refused and he really got mad at me."

David Olivetti thought Visentini was the wrong choice because he did not have the necessary background. He was being advised by Valletta and his group, "and they were telling him how they wanted Olivetti run, i.e., *out* of the running. One more example of people in positions of power at Olivetti who were working to destroy it."

&

By following the careers of other executives after the interventionist group had done its worst, one discovers some interesting anomalies. Men who had been identified with the objectives of Adriano, Roberto, and Mario faded away. Others saw their fortunes rise after Fiat took control. Gianluigi Gabetti was a young bank executive who had been hired by Adriano in 1959 some months before his death and quickly became a major financial advisor. During an interview in Turin, Dott. Gabetti gave a lengthy description of his hiring and the events leading up to the Underwood sale. He was at pains to explain that he had not advised Olivetti to buy because, he said, he thought it should be Adriano's decision. He went on to be appointed president of the Olivetti Corporation of America by Visentini in 1965. In 1971 he became director-general of IFIL, the Agnelli family's investment company, and subsequently became a close and longtime financial advisor to Gianni Agnelli.

One would have expected Ottorino Beltrami, a member of Adriano's inner circle, to be fired. The curious fact was that he, who was in charge of the Electronics Division before the coup, was not only *not* fired but awarded the same job after it was sold to General Electric, supposedly to further its prospects but actually to shut it down, as Perotto discovered. When Visentini fired Roberto Olivetti in 1971, Beltrami received an even bigger promotion. He was hired to replace his former boss. Not only must this have been humiliating for Adriano's son, but he then had to watch the ironically named "Comandante" systematically run the company into debt until it, too, was almost destroyed.

The evidence speaks for itself. Beltrami was continuing a clear mandate to shut down Olivetti or at least prevent it from ever colliding with American interests again. Beltrami had worked for Italian military intelligence, probably the Nazis too, and may have known all about Gladio and its Mafia links. Then he worked for the Marshall Plan, aka the CIA. What double role had this affable, noncommittal former spy taken in the Machiavellian games being played? What did he tell Adriano on that last fateful train trip that changed his mood from one of buoyant optimism to foreboding?

When Carlo De Benedetti replaced Beltrami in 1978, "he was the company's last chance. If he hadn't been willing to invest they would probably have had to go into bankruptcy," David Olivetti said. Seven years after Beltrami took over as managing director, the company was losing $10 million a month, and its debts were over $1 billion. "Productivity and morale at its plants had hit rock bottom, and aggressive competitors from the United States and Japan were sweeping Olivetti's mechanical typewriters and calculating machines from the marketplace," John Tagliabue wrote in *The New York Times.* Six years after De Benedetti assumed control, the company had successfully introduced a new line of electric typewriters, it was back in the running as a computer company and its sales of $2.3 billion had restored its position as the second largest office machine company in Europe after IBM.

One of De Benedetti's first moves as chief executive was to invite Roberto, even though he had no money to invest, to join his new board. "He was treated very badly by his family," De Benedetti said. "The family tried to build an image around him that he was not competent, he couldn't cope. It is hard to know just why." Dott. Gabetti shared the family's view. "Roberto Olivetti was weak," he said. Asked what he meant by that, Gabetti looked surprised, as if the answer were self-evident. "Because he had no money," he said.

ↄ

How did Adriano feel about his only son? One does not know but a hint of his attitude is suggested in an interview with him in 1959 that probably took place shortly before his death. He said Roberto was a "mix" between himself and Gino Martinoli; "add us together and stir, and out comes Roberto." He also called his son "brimming with cast-iron energy" and "vitality." One wonders how much this description of Roberto, as some kind of extension of himself, affected their relationship. Roberto's daughter, Desire, concluded that her father suffered from the hopeless goal of ever living up to expectations. At the same time Roberto felt a personal obligation to protect and defend his father's achievements, if only for all the people whose livelihoods depended upon it. He spent most of 1963 traveling around Europe in an effort to build a multinational consortium—it would be called La Elettronica Europa—in order to save the electronics division. No luck. He failed again to raise loans with the banks, even Swiss ones. So did Dino, who was attempting to do the same in New York.

"Roberto may have been a man alone," Roberto Saibene wrote, "but in those years he was at least not short of self-confidence. . . . [H]e drove fast, skied hard and sailed solo from the Italian coast to Corsica in a dinghy." Saibene might have added that Roberto was also something of an architect manqué; he favored the Bauhaus, admired everything Sottsass invented, and if he needed a chair, Roberto Olivetti would design one himself.

In 1966, as soon as he became managing director, he turned his attention to Underwood, where he was free to act. Almost his first decision was to move the plant from its antiquated headquarters to a new site in Hartford. He wanted the right architect and turned to Louis Kahn. As has been noted, the company was also known for commissioning famous men to design its work spaces and Kahn was delighted to accept. The architect was faced with a requirement to make the factory as open as possible, in order to meet whatever modifications or change of direction might be needed.

It seemed like a tall order. But Frank Lloyd Wright had successfully demonstrated that it could be done with the Larkin Company building, a single open space surrounded by balconies, and with his groundbreaking design for the Johnson Wax building in 1939. For the latter he designed a large, windowless rectangle decorated, at the roofline, with a frieze of glass tubing that admitted light, but no view. The same glass tubing was used in the roof. Inside, the central work space was interspersed with rows of slender concrete columns sometimes called mushroom, or lilypad, in shape. Louis Kahn built a series of "units" in a cluster, each of them looking like "a square dish with clipped corners perched on top of relatively thin columns." The interiors were vast and mostly empty, but mimicked the lilypad design that had given the Johnson Wax building its particular distinction. Translucent skylights in the roof looked exquisite but, since the walls were 30 feet high, presented practical cleaning problems once snow, ice, and city dirt had done their work. But, for the moment, these elegantly simple spaces with their uncluttered facades were the blank canvases for their future occupants, and the words "olivetti underwood," all in lower case.

The next challenge for the architect in Roberto Olivetti was the need in Ivrea to provide small apartments for company executives and the many visitors for whom there was little accommodation available. He and his architects, Roberto Gabetti and Aimaro Oreglia d'Isola, settled on a gently sloping field and inserted into the slope a crescent-shaped row of apartments and maisonettes. The tenants had a clear, unfettered view of fields and trees; the backs of the apartments were actually inside the bank, with access for cars behind them and a large walkway overhead that acted as a roof. The project had the prosaic title of West Residential Unit but was quickly renamed by the town as "Talponia," or Mole City. The effect of glass facades facing a field of flowers is quite stunning. It brings to mind many of Wright's

The pedestrian walkway curving around on top of "Mole City" in 1971

designs that have been integrated into sloping sites that take advantage of similar views, along with his dictum that a building should not command and dominate the landscape, but be enfolded within it.

The least successful of Olivetti's building projects in those years was La Serra (1967–1975). This was a multipurpose complex of hotel rooms, plus a cultural center (cinema, exhibition space, conference hall), plus a sports center: swimming pool, sauna and gym, plus a restaurant and bar, and so on. One would be prepared for such a building to be an unruly muddle; and to its already daunting complexities was added the further challenge of being able to get from anywhere in the building to anywhere else. Plans were continually being amended to make this possible. Then there was the truly awful discovery that the site chosen, right in the center of town, turned out to have archaeological remains. So further redesigns were necessary to protect what was underneath. If the young Venetian team of architects, Iginio Capponi and Pietro Mainardis, had meant to put one in mind of a typewriter, the necessity for continual emendations has blurred the original intent. No one seemed to have questioned a massive concrete and glass structure in the center of all those tile roofs, green shutters, and soft, burnt-sienna facades.

La Serra was obviously meant as a well-meaning gift to the town, but "people thought of [these] buildings as alien spaceships," Luca Marrighini, an architect, said in the Elea 9003 documentary. The town has dealt with the building in its own way. It is now empty. The glass windows are streaked and the dingy white walls are covered with graffiti. The locals walk around it as if it doesn't exist.

⌘

After Roberto Olivetti left the company, one of his achievements was to create what has now become a flourishing enterprise, the Fondazione Adriano Olivetti in Rome. Even then, his presidency was opposed by his aunt Silvia but for once she was overruled. In 1977, he became general manager of a financial company set up to promote industrial development in the south. He ran briefly for office (and failed); his interests in publishing, art, and sailing continued, as did his interest in the theories of Freud and Jung. He formed a stable union with Elisa Bucci-Casari that lasted until he died. Carlo De Benedetti, who saw them often, thought she was the right person for him. "Roberto was very shy," he said. "She helped him a lot." She initiated lively dinners and parties with friends in the worlds of Roman politics and the arts that they inhabited. She loved to go sailing. She plunged into things, the way Paola did, another plus for their relationship. De Benedetti said, "She took him around and cared for him until he died." They were married quietly a couple of years before that happened.

His death came about in the spring of 1985. In later years Roberto seemed to have come to terms with events. But a distinct coolness, perhaps inevitable, had sprung up between him and certain members of his family. Some people thought he was bitter, but not his daughter. She described his mood as a persistent sadness or melancholy, "a kind of shadow in front of him." He had always smoked and now he began to smoke heavily; two packs a day. Desire said, "I pleaded with him to stop. I was on my knees. I said, 'In a week we go skiing. Can you ration yourself to ten cigarettes a day?'" He tried, but "of course" a week was not long enough and the heavy smoking began again.

Desire recalled that her father had been aware that something was wrong physically but, stoic that he was, he ignored it. "Then one day when he was in bed he turned on his side and could feel something." X-rays were taken and doctors discovered "a massive growth inside one of his lungs." It was removed in 1983. "He lost more than half of one lung." Desire was told that her father had two more years to live. She was about to start classes in the University of Pennsylvania School of Architecture, but canceled them to be with her father.

During the next two years Roberto was in and out of the hospital for operations and chemotherapy, but nothing helped. He became pale and thin as the cancer reached his throat—he could neither swallow or, at times, even talk. In his final days he was characteristically telling visitors that he was recovering and would be leaving the hospital the next day. He died in his sleep, April 25, 1985.

Inge Feltrinelli thought that, in the 1950s and 1960s, few people were on Roberto's level. "In a country that is not very scrupulous about political morality, Roberto was intact, consistent. He could be

*Bruno Caruso painted this faux-naïf work to celebrate Roberto
Olivetti's skill as a sailor in saving the lives of a family boating party
that was hit by a violent storm off the Mediterranean coast, 1963*

an excellent minister. But it was not to be; he was too honest, too kind, too smart. He was a man of the world, charming and personable, but at the same time authentic, morally rigorous, not to say puritannical, as well as a man of culture." He had acumen, foresight and he was persistent. There would have been no P101, no breakthrough in the world of personal computing, without Roberto Olivetti, who persuaded his father to begin research, who found Mario Tchou and gave him his head, funded the laboratory, guided its research, lied on its behalf, and kept the critics in the company at bay. How could anyone consider him a failure? He was a star, Inge Feltrinelli wrote. But the effort to prevent the electronics division from being shut down and, finally, to be forced to watch the company being torn apart, proved too much for him. Here was yet another victim of the Cold War. When he died, he was fifty-seven years old.

ↄ

No one lives in the Convent nowadays. The windows are blank and the stucco is peeling. A future theatre that would have occupied the site opposite it remains unbuilt. The tennis courts nearby, surrounded by wire fences, are not in use and the wild grasses have arrived. A great emptiness surrounds the huge glass walls of factories on the via Jervis. These once formed the backdrop to photographs of arriving workers, with their bicycles, jammed shoulder to shoulder across the road. The heavy white iron gates that served as the entrance to Camillo's red-brick buildings are locked and unmanned. The kiosk beside the entrance that once sold *La Stampa* and many other newspapers is a sad reminder of these years of neglect, its metal shutter long since pulled down for the last time.

These vast glass facades, factories with rusting supporting beams, their sills covered with dust and litter, bear mute witness to the work that once went on behind them. There is an occasional parked car. But few drive down that stretch of the via Jervis, named for Guglielmo "Willy" Jervis, an Olivetti engineer who fought with the partisan wing

Via Jervis today

of Giustizia e Libertà, was captured by the SS, tortured, and shot in August 1944. What is left behind conveys a feeling as desolate and disorienting as a dream.

The same persistent, curiously unsettling atmosphere hangs over the public cemetery in Ivrea, which, to judge from its fresh floral offerings, is well remembered and often frequented. Yet even on a sunny morning, with the foothills of the Alps in the gray-blue distance, there is still the same feeling of eerie unreality. As in Böcklin's famous painting *The Isle of the Dead*, with its cypresses, columns and pediments, arches, and heavy stone plaques, the atmosphere is sombre. One walks down alleys far too wide, past row on row of boxes containing the ashes of former citizens, all of them bearing photographs of the departed, with a sense of loss and foreboding.

The memorial garden to Adriano Olivetti is hard to find; it is tucked away in a small grove, planted with oak leaf hydrangeas, roses of Sharon, and evergreens, most of them in need of pruning. A flowering tree has cast its small pink petals over a simple wooden cross, with Adriano's name carved on the right arm and Grazia's on the other. Opposite this is a ten-foot statue of St. Francis in black basalt, wearing a monk's hood and curiously Eastern features; he could be a Buddha. Below his waistline is a mosaic bas-relief in tiny glints of blue, gold, and orange. The eyes of a central figure are staring, the mouth open in a scream of disbelief; mourners behind him wear the same anguished expression. Sunlight does not penetrate this dark and mossy spot. It is still, solitary, and appears seldom visited.

The lives of all three men—Adriano, Camillo, Roberto—are unfinished, because the story is unfinished; cut off and dispensed with, the work deliberately obstructed, confounded, or stolen, hedged around with hidden obstacles, thwarting what was begun with so much care, energy, and goodwill. As for the company, having spurted fitfully, changed direction, its shape, sold off its divisions, and dispersed its workers, that, too, has left nothing but a footnote in the industrial history books. It, too, has slid into oblivion.

Perhaps not quite. "At least we can talk about the Olivetti family now, what Olivetti has done," Dott. Nerio Nesi said. "For many years the name of Adriano Olivetti and all those who worked for him, had to be erased by the various governments without party distinction, erased from the most important order of merit in the workplace. We were told, 'An experiment like Adriano Olivetti's *must never* happen again.' This was the general approach. This was the problem. The fact that it is being talked about ever more frequently is a good sign."

Another piece of good news is the inclusion of "Olivetti City" into that select roster of World Heritage sites deemed worthy of preservation by the United Nations Educational, Scientific, and Cultural Organization. In July 2018, UNESCO cited the specific level of architectural achievement that the site showed. The model social project expressed "a modern vision of the relationship between industrial production and architecture."

Beniamino de' Liguori Carino, director of the Olivetti Foundation, thought that the achievement went further. Olivetti City embodied "a human heritage." Its success was in creating a positive, even ideal relationship between man and his work; as Massimo Fichera expressed it, "The factory not only as a producer of goods. The factory as a producer of good."

ɕʒ

In retrospect it is clear enough to understand the antipathy that greeted Adriano Olivetti's ideas, as well as those of another great social reformer, John Ruskin (1843–60), who was the first to see in architecture a reflection of social dis-ease. Kenneth Clark wrote that to understand the horror that greeted Ruskin's first essays on capitalism and its deleterious effects, written in the 1850s, one must look at "orthodox political economy." This was "the theology of the only effective 19th-century religion . . . It disguised, or justified, the fact that a few people were making more money than they required by exploiting the poor. To question this theology was a threat to the peace of mind of all property owners."

Gandhi often said that Ruskin was one of the principal influences on his life, and Tolstoy thought him one of the greatest social reformers of his time. Clark wrote that "[n]ot only the direction but the whole style of attack of the early English socialists is unmistakeably

Now deserted, the La Serra building's paint is peeling, its
glass facades are grimy, and graffiti has arrived, 2016.

Ruskinian." One of them was William Morris (1834–96), an English artist, author, socialist, textile designer, and businessman closely linked to the Pre-Raphaelite Brotherhood and the English Arts and Crafts Movement. Like Adriano Olivetti, Morris was the son of well-to-do parents and there are further parallels having to do with the quixotic, painful, and ultimately idiosyncratic journeys both men took through life.

Their temperaments are remarkably similar as well. Henry James described Morris as "short, burly, corpulent, very careless and unfinished in his dress . . . He has a . . . nervous, restless manner and a perfectly unaffected and businesslike way" of speaking that has distinct parallels with the descriptions of Olivetti. Like Adriano's, Morris's address was "wonderfully to the point and remarkable for clear good sense." J. W. Mackail, who was Morris's first biographer, commented on his subject's many admirable traits. He was "industrious, honest and fair-minded," and exhibited "a strong sense of family responsibility"; so did Adriano. Both men, while capable of lasting friendships and deep affection, were, ultimately, emotionally inaccessible. Mackail wrote that Morris did not allow people to "penetrate to the central part of him"; one could have said the same about Adriano.

The resemblances continue with the beliefs deeply held by both men. Both were lifelong revolutionaries, and anti-imperialist in outlook. For them, state socialism and centralized control were anathema; they sought for localized control. As Morris wrote, "what I mean by socialism is a condition of society in which there would be neither rich nor poor, neither master nor master's man, neither idle nor overworked . . . in which all men would be living in equality of condition . . . and with the full consciousness that harm to one would mean harm to all—the realisation at last of the meaning of the word *commonwealth*."

One can discern, in Olivetti's lifelong emphasis on the need to link art and technology, Morris's most famous saying, "If you want a golden rule that will fit everybody, this is it: Have nothing in your houses that you do not know to be useful, or believe to be beautiful." Both viewed machines as important only insofar as they remained the servants of mankind. The idea that machines might ever become "our masters and not our servants" horrified both men, and their likely reactions to the spectre of Artificial Intelligence goes without saying. Ultimately, the message as expounded by Olivetti as well as by Morris was "the abandonment of capitalism itself and its replacement by more equitable, human structures," as Fiona MacCarthy wrote in *William Morris: A Life for Our Time*. Morris married "the tradition of socialism as a critique of political economy" with the "tradition of antic anti-industrialism"; so did Olivetti. Such men come along only once or twice a century, being both of their time and beyond it; they were "time-travelers." Again, as Morris once explained it, "If others can see it as I have seen it, then it may be called a vision rather than a dream." Something about Adriano Olivetti's perennially youthful and calm certainty could be quite terrifying. The same can be said of the expression in his clear, childlike, and penetrating eyes.

Acknowledgments

As I have mentioned elsewhere, the book had its origins in a case of idle curiosity. As I discovered a few of Roberto Olivetti's obituaries in English I was struck, not only by the fact that he was a much more important figure in the company than I knew, but that he died at the relatively early age of fifty-seven. There was no mention of the cause of death. But now I was interested.

The foundation in Rome named for his father, Adriano Olivetti, could not tell me either. What they did was to put me in touch with Desire Olivetti, Roberto's only child, a daughter of about my own daughter's age. (Pronounced as desire and not in the French fashion, as Desirée.) All of a sudden a host of new ideas presented themselves. Almost my first question to her was, what did her father die of? Since I was sure he must have "wrapped himself around a tree," as I said. "We all thought that," she responded with remarkable frankness, before describing the illness that had killed him. She recounted the financial crisis following the death of Adriano in 1960 and the business crisis that, it was said, resulted from her father's inability to act. Family members thought Roberto was indecisive and lacking in the persuasive powers and inner strengths his father had displayed. A recent Italian TV semidocumentary about Adriano, *La forza di un sogno,* was the first to launch the thesis that CIA influences had been at work, as well as IBM. These forces stopped the company as it attempted to move from a mechanical one and into electronics. But the theory had never been proposed before. This was the first of many conversations and

I owe her more than thanks, which are heartfelt, for setting me on a study that could be called the rise and fall of a great institution.

Roberto had American cousins, the sons of Dino, Adriano's youngest brother, who, in the postwar years headed the North American division. He and his wife, Posy, moved to Connecticut, where they raised their bilingual children. Philip became a crack salesman and was now living in New York, so in the summer of 2015 he and I had the first of what turned out to be many interviews. Biography requires, as a rule, the slow accretion of detail and consequently there are few "Eureka!" moments. But in this case, one revelation after another arrived during the course of our first afternoon. Not only had the company developed and successfully launched the world's first desktop computer, known as the P101. But, at the moment when it was unveiled at the New York World's Fair in 1965, forces behind the scenes had already shut down the company's electronics division, ensuring that Olivetti's first brilliant electronics innovation would be its last. As for Adriano's death on a train at the early age of fifty-eight—always described as a heart attack—"My father used to think he had been murdered," Philip Olivetti said. He continued, "I made a big effort to stay out of it." Just the same, there were reasons to think it might be true. Then there was the curious death of Adriano's chief engineer, Mario Tchou, in a flashy American car on a rainy autoroute a year later . . . The story had not yet been told, and it was high time someone did. I flew to Rome in the spring of 2016 and spent several weeks interviewing there, as well as at Ivrea, the company's former headquarters, Florence, Turin, and Milan.

Philip's brother David was another absolutely vital source of information about the company and Adriano's family, which included five brothers and sisters. He, his wife, Lynn, and their children settled in Ivrea, where his grandfather Camillo first set up his business as a typewriter company in 1908. Like many members of the Olivetti dynasty, he and his children were bilingual—Lynn is an American—giving them a unique perspective on both cultures—political, economic, and social. He became an indispensable guide, showing me the Convent where Camillo first set up shop, the church, the schools and workers housing Camillo and Adriano built for their workers, the factories, offices, canteens, and libraries, all of it set in rolling meadows and bosky arbors. A fine photographer himself, David has put together a detailed portrait of the Olivetti family beginning with Camillo's parents in 1859 to the present day. His family album also contains a meticulous genealogical index of Camillo's descendants. David Olivetti and his son Matteo, an architect and Olivetti authority, gave me introductions to family friends and former employees of the company who have been extraordinarily helpful themselves. Matteo's filmed interviews over many years of former Olivetti executives, now deceased, provide important clues to what was happening behind the scenes as the company was taken over and its electronic division shut down.

Among other family members who were kind enough to help I am indebted to Alfonso Merlo, Magda Olivetti and her daughter Tatiana Jaksic, Paolo Marselli, Mimmina Galardi, Dott. Alberto Galardi and Annalisa Galardi, Anna Olivetti, her husband, Antonello Nuzzo, and their daughter Paola Nuzzo, Gregorio Cappa, and Maurizio Galletti. I was particularly pleased to meet Roberto's sister Lidia, who was living with her second husband, the painter Bruno Caruso, when I was invited to have lunch with them in their picturesque country house on the via Antica in Rome. Lidia, with her quiet charm and long memories, was particularly helpful about her girlhood living in Fiesole during the war. I was also thrilled to meet her daughter Albertina Soavi, who is an art restorer, and whose kind help has been invaluable, as well as Lidia's son Michele Soavi, the mind behind the 2013 RAI-TV program on the life of Adriano Olivetti. Lidia died at the age of eighty-nine in May 2018. Yet another authority on the subject, Anna Nogara, the actress and former wife of Roberto Olivetti, Desire's mother, has been endlessly helpful and hospitable. Her particular perspective has been invaluable. I also want to thank Elisa Bucci-Casari, who invited Desire and myself to visit the apartment in Rome she shared with Roberto, and where I also met her daughter, Letizia Maraini.

This book could not have been written without the generous help and assistance of the Fondazione Adriano Olivetti in Rome, its director, Beniamino de' Liguori Carino, and immensely knowledgable assistant, Francesca Limana. The foundation, which was created by

Roberto Olivetti in the 1960s, is a major repository for books, articles, and unpublished materials about its subject as well as publisher, under the banner of Edizioni di Comunità, of a handsome list of books by and about Adriano Olivetti that includes a biography by Valerio Ochetto, the first to be published, in 1985, which is still in print. The publishing house, under Beniamino's guidance, recently published an indispensable guide to Ivrea's many links with Olivetti, from a historical, industrial, and sociological perspective.

When the foundation learned I was making a visit, Francesca Limana spent weeks arranging interviews with engineers, academics, artists, authors, politicians; in fact, anyone who had memories of Adriano and Roberto, and could put unanswered questions into perspective. I owe them both an enormous debt, but particularly Francesca, who drove me around Rome with a panache and daring that alternately terrified and amazed me. The foundation also facilitated my study in Ivrea at the Archivio Storico Olivetti, which also contains extensive business and family records. This archive, through some quirk of fate, no longer belongs to the family but is owned by companies that bought out Olivetti over the years, and provides access by appointment to researchers. I am indebted to Lucia Alberton, who bore the arduous process of retrieving, then returning, endless boxes of Adriano's letters, with patience and good humor.

Shortly after I began working in earnest on this project, in November 2015, I contacted Glen Miller, a spokesman in the Central Intelligence Agency, requesting an interview with the agency's historian, David Robarge. By then I knew that Adriano Olivetti had met the agency's former director, Allen Dulles, in Switzerland during World War II and had carried on an active correspondence with him afterward, most of it to do with a new political party the former was organizing and for which he hoped to elicit American support, if not actual funding. A few of these letters had been released under the Freedom of Information Act. I hoped to get that relationship into better perspective, and also be allowed to fill in what seemed to be some missing gaps. Somewhat to my surprise my request was denied, the reason given being that the issue was still "sensitive." How could that be the case, I asked, seventy years later?

However there was no shortage of good counsel from expert researchers and authors who had been presented with the same obstacles. I owe a debt of gratitude to Jeff Goldberg, who had acted as chief researcher for *Cold Warrior,* the biography of James Jesus Angleton by Tom Mangold published in 1991. He was enormously helpful in identifying collections of archives where some reference to Olivetti might be found. I am also indebted to Bruno Amoroso, whose collection of essays, *Capitalismo Predatore,* I had already read and whom I interviewed in Rome. My thanks also go to Dr. Lucia Wolf of the Library of Congress European Division, who was the first to suggest I might also consider the role of Italian businesses and their possible interest in handicapping the Olivetti Company at the height of the Cold War. Dr. John F. Fox Jr., historian for the Federal Bureau of Investigation, was particularly helpful in guiding me through the massive testimony made public by the Church Senate Committee hearings on the role of the CIA during the Cold War and the techniques of covert action then being used. William Blum, author of a ground-breaking study of CIA assassinations at the same period, *Killing Hope,* and particularly the role played by the U.S. government during the Italian elections of 1948, was another vital guide. I lunched in Rome with Philip Willan, who had published *Puppet Masters: The Political Use of Terrorism in Italy* in 1991. Philip and I shared the same suspicions about Operation Gladio, and the secret armies put in place in Italy and elsewhere by the North American Treaty Organization in 1947. Unbelievably, those armies were officially denied for decades; it was not until 1990 that the Italian prime minister conceded that such an army had existed in Italy since 1958, and that, under U.S. control, Gladio had aided and abetted right-wing terrorism in Italy for decades. The CIA, it was suspected, had secretly used Gladio to bring about the capture and murder of Italian prime minister Aldo Moro in 1978.

Like Philip, Jefferson Morley, a former colleague of mine at *The Washington Post,* who lectures frequently on the CIA, shared similar concerns and has also written extensively on the subject. His new biography of Angleton, *The Ghost,* was published in 2017 to glowing reviews. Following Jeff Goldberg's advice I looked up the Association of Former Intelligence Officers in Falls Church, Virginia, and was put in touch with S. Eugene Poteat, one of its former presidents. Poteat is the author of a study dealing with espionage after the Cold War. I also spent an interesting afternoon with Gene and his friend, the psychiatrist Bill Anderson. We pondered the significance of body language—Kim Philby, the British traitor, being a case in point—and the way his questioners might have looked for clues, not in what he said, but in the way he was wringing his hands.

Philby is a special interest of Michael Holzman, another wanderer in the maze of misinformation that constitutes the spy's world. His own biography, *James Jesus Angleton, the CIA, & the Craft of Counterintelligence,* is quoted at length in my text, perhaps because his elegance of style and gift of storytelling are irresistible. I am also a great admirer of Dr. Darren Tromblay, whose ability to summarize is formidable. Dr. Tromblay is an intelligence strategy consultant and author of the recent book, *Political Influence Operations: How Foreign Actors Seek to Shape U.S. Policy Making* (2018). There are many other authors whose work has been an inspiration: Daniele Ganser, author of the seminal study, *NATO's Secret Armies* (2005); Alan Friedman's brilliant analysis, *Agnelli and the Network of Italian Power* (1988); Paul Williams, *Operation Gladio* (2015); and John Hooper's *The Italians* (2015). I would also like to pay tribute to a collection of oral histories by former State Department employees. A search through this collection for men and women stationed in Italy during the Cold War turned up a number of unguarded accounts, which shed surprising light on the attitudes of Foreign Service staff in Italy at this time. Just what did they reveal? It turns out, quite a bit.

My impressions of Adriano's character have been put together from many sources, but particularly the long interview I had with Dott. Franco Ferrarotti, author, professor, sociologist, and politician, during the spring of 2016. Dott. Ferrarotti, who met Adriano at the end of World War II, is particularly well placed to analyze the unique qualities and contradictions of this complex personality and I am very much in his debt. So are Dott. Nerio Nesi and his wife, Dr. Patrizia Presbitero, who invited David Olivetti and myself to lunch in their exquisite villa on the hills of Turin. Dr. Nesi, a banker and longtime leader of the Italian Socialist Party, is frequently invited to take part in Italian TV and other documentaries. He has recently published his own reminiscence about the Olivetti family. His testimony is especially precious now that there are so few left alive who remember Camillo and Adriano. To Dr. Nesi and his wife in particular, I offer my warmest thanks.

So many other people have helped with queries, information, documents, photographs, and interviews and I am indebted to one and all: Dan Amory, Raymond V. Arnaudo, Ing. Roberto Battegazzorre, Mario Bellini and Elena Marco, Franco Bernini, Prof.Ing. Ugo Bilardo, Dr. Ing. Lucio L. Borriello, Susannah Brooks, Gregorio Cappa, Giacomo de' Liguori Carino, Domenico de' Liguori Carino, Matilde Cartoni, Vinton Cerf, Dr. Paul Ceruzzi, Paul Colby, Sally Shelton Colby, Carl Colby, Dr. Bruce Cole, Furio Colombo, Roberta Colombo, Dott. Pietro Contadini, Brenda Cronin, Dr. Mary Elizabeth Curry of the National Archives, David Holbrook of the Dwight D. Eisenhower Presidential Archives, Ing. Carlo De Benedetti, Lorenzo Enriques, Mariangelo Michieletto of the Archivio Nazionale Cinema d'Impresa, Dan Esperman, Anna Foa, Jeanluigi Gabetti, Amb. Richard N. Gardner, Ing. Giampieri Garelli, Gastone Garziera, Dott.ssa Giuliana Gemelli, Milton Gendel, Paola Giovannozzi, Antonio Giusti, Regis D. Heitchue, Grazia Goseco of the Embassy of Italy; Dott. Renato Miracco, cultural attaché of the Embassy of Italy, David E. King, senior engineer of MEA Forensic Engineers and Scientists, Luci Koëchlin, Giorgio la Malfa, Mike Ledeen, Gabriella Lorenzotti, Amb. Frank Loy, Paolo Lupatelli, Dott. Paolo Marselli, Jeffrey Kozak of the Marshall Plan Foundation, Alfonso Merlo, L. G. "Nick" Modigliani, Arthur Molella, director emeritus of the Lemelson Center, Smithsonian Institution, Cristina Monet Zilka, Elisa Montessori, Lorraine Motta, Silvia Napolitano, Jerry Nedilsky, Tim Nenninger of the National Archives, Marybeth Ihle of the New-York Historical Society, David Patrick Columbia, *New York Social Diary,* Barbara Negri Opper, Antonio d'Orrico, managing editor, *Corriere de la Sera,* Lorenza Pampaloni, John and Lynette Pearson, Pier Paolo Perotto, Giuseppe Petronzi of the Embassy of Italy, Walter Pincus, Murray Pollinger, Meo Ponte, John Prados, Mme Andrea Pyrothe of the

Archives d'Aigle, Counsellor Giuseppe Rao, Noreen Rainier, James Reston Jr., Carlo Ripa, Betty Sams, Clarice Smith, Luzia Furrer, vice consul, Embassy of Switzerland, M. Gilbert Coutaz of the Association Vaudoise des Archivistes, Corinne Brelaz of the Archives Cantonales Vaudoises, Dr. Peter Fleer of the Schweizerisches Bundesarchiv Bar, Abteilnng Informationszugang, Alberto Vitale, Prof. Thayer Watkins, Tim Weiner, Kay Wilson, and Wim de Wit.

As I was writing, I became more and more aware of the complex nature of the detective story that was unfolding. It was clear that the usual explanations could not be made to fit all that was happening. But just how it did happen and who was involved—whether the CIA at Rome station had engineered Olivetti's fall, how much Gladio forces came into play, and what role politics, colored by business pressures, had contributed—these are issues that, perhaps, time will one day answer. In the daily struggle I would like to thank my husband, Thomas Beveridge, whose talent for seeing issues clearly is matched only by his patience for listening to the same arguments one time too many. As always, I am infinitely grateful for the day I was introduced to my incomparable editor, Victoria Wilson, more than twenty-five years ago. I am further indebted to my superb agents, Cullen Stanley and Lynn Nesbit, who have never ceased to provide emotional support. There is one person whom, alas, I cannot thank personally. Roberto Olivetti's silent and desperate struggle spurred me on, as did his blithe and indomitable spirit. He is to thank for starting it all.

Meryle Secrest
Washington, D.C.

Notes

Abbreviations

BAR — *The Italians,* Luigi Barzini, 1964.

BRITISH — *A British Boy in Fascist Italy,* Peter Ghiringhelli, 2010.

CA — *Gli Olivetti,* Bruno Caizzi, 1962.

CENTI — "Cinquantemila," an Adriano Olivetti chronology, 2015.

CLARK — *Modern Italy: 1871–1995,* Martin Clark, 1998.

ELEA — *Elea 9003:* film documentary by Pietro Contadini, 2009.

ELECT — "Memories and Anecdotes of Olivetti Electronic Research Laboratories," Giuseppe Calogero, August 2011.

FRIED — *Agnelli and the Network of Italian Power,* Alan Friedman, 1988.

GAN — *NATO's Secret Armies: Operation Gladio and Terrorism in Western Europe,* Daniele Ganser, 2005.

GINZ — *Family Sayings,* Natalia Ginzburg, 1963.

HAR — Harvard Business School report on Ing. C. Olivetti & C, 1967.

HOOP — *The Italians,* John Hooper, 2015.

K&M — *Invented Edens: Techno-Cities of the Twentieth Century,* Robert K. Kargon and Arthur P. Molella, 2008.

LA — *Lettere Americane,* Camillo Olivetti, 1999.

LEW — *Naples '44,* Norman Lewis, 2005.

OCH — *Adriano Olivetti: La Biografia,* Valerio Ochetto, 1985.

ORAL — Association for Diplomatic Studies and Training, Foreign Affairs Oral History Project, various dates.

PER — Pier Giorgio Perotto: *P101: Quando l'Italia Invento il Personal Computer?* undated.

POL — *A History of Contemporary Italy: Society and Politics, 1943–1988,* Paul Ginsborg, 2003.

RADICE — *Ettore Sottsass: A Critical Biography,* Barbara Radice, 1993.

SEC — *Frank Lloyd Wright,* Meryle Secrest, 1992.

TAL — *The Devil's Chessboard: Allen Dulles, the CIA, and the Rise of America's Secret Government,* David Talbot, 2015.

WEI — *Legacy of Ashes: The History of the CIA,* Tim Weiner, 2008.

Chapter 1: Oranges

9 honor is satisfied: Interview with David Olivetti.
11 the *Tocco Olivetti: Fortune,* Ibid., p. 139.
12 considered a legend: Ibid., p. 101.
13 ". . . It didn't seem important": Milton Gendel to author.
13 This particular Saturday: weatheraprtk.com
13 "I was the last . . .": Franco Ferrarotti to author.
14 "I have never seen him . . .": Posy Olivetti, Interview with Matteo Olivetti.
14 "nervous, even agitated": *La Stampa,* February 29, 1960.
16 some important documents: *Il Tempo,* Rome, March 10, 1960.

Chapter 2: "A Clear Head and a Nimble Leg"

17 "the moving waters . . .": "Bright Star," sonnet by John Keats, 1819.
19 "perpetually confused . . .": Notes, p. 254, LA.
19 "My poor mother . . .": Ibid., p. 197.
19 allowed to run wild: CA, p. 10.
20 he or she would eat less: *Society and Politics in the Age of the Risorgimento,* ed., John A. Davis and Paul Ginsborg, p. 170.
22 "[T]he chief fighters . . . ": CLARK, p. 112.
22 its first Socialist member: *Adriano Olivetti: A "Socialist" Industrialist in Postwar Italy,* by Claudio Nasini, chapter, "Italy and the Bourgeoisie," p. 77.
22 "a subversive individual . . .": Ibid., p. 85.
23 ". . . two hundred well-armed men . . .": Preface, LA.
24 ". . . the ugliest period . . .": LA, p. 21.
25 "I don't waste my time . . .": Ibid.
25 ". . . an anti-ecclesiastical bias . . .": HOOP, p. 124.
27 ". . . thinking of you": LA, n.d.
27 bound to succeed: Letter from A. A. Barlow, April 25, 1899.
27 ". . . were never divulged": P101, p. 16.
28 no one wanted it. HAR, AM-P, p. 224.
29 "The M1 was offered . . .": *Olivetti Builds,* by Patricia Bonifazio and Paolo Scrivano, p. 12.

Chapter 3: The Convent

32 ". . . the truly marvelous progress . . .": HAR, AM-P, p. 224.
33 ". . . very confused ideas": GINZ, p. 61.
33 fear of brucellosis: Ibid., p. 62.
33 to stop scrounging: Ibid.
34 "acutely sensitive": OCH, 23.
35 ". . . a very simple life . . .": November 18, 1923.
36 ". . . how good and kind . . .": GINZ, 61.
36 She didn't like pumpkins: Interview.
37 ". . . a very caring person": Ibid.

37 ". . . his declining years . . .": HOOP, p. 160.

37 "I was imprisoned . . .": CA, p. 132.

37 ". . . the accursed conditions . . .": *C. R. Ashbee,* by Alan Crawford, p. 419.

39 Silvia was secretive: OCH, p. 27.

40 ". . . many willing soldiers": CA, p. 129 (c. April 1918).

40 ". . . cold and dreamy": GINZ, p. 60.

42 "bad-tempered door slamming": Ibid., p. 54.

42 ". . . like a lioness": Ibid., p. 55.

42 Armistice was declared: On November 9, 1918.

43 It ran for a little more than a year: From August 1919 to October 1920.

45 she could not marry him: OCH, p. 34.

45 "something contemptible . . .": GINZ, pp. 56–57.

46 her brilliant, doomed life: Ibid.

Chapter 4: Enter Adriano

49 ". . . someone coughing . . .": GINZ, p. 68.

49 ". . . huge as a bear . . .": Ibid.

50 ". . . I saw that look . . ."; Adriano had to disappear: Ibid., pp. 68–71.

50 ". . . a visible symbol of aspiration . . .": *New York, 1948,* by E. B. White, p. 31.

51 The skyscrapers, he wrote, were truly grand: This and other comments are taken from letters Adriano wrote to his family on visiting New York for the first time in 1925 and 1926.

51 "idiot games": Ibid.

52 to the Louvre: Ibid.

52 "loss of five cars": Ibid.

52 "This childish America": Ibid.

53 awkward and stupid: From a letter to Grazia, his future wife, undated, c. 1949.

53 ". . . apparently superior . . .": Letter, January 31, 1927.

54 fast and simple: Letters dated March 10, 1927, and February 16, 1927.

56 He would wake her up: Ibid.

57 ". . . the spirit of independence . . .": CLARK, p. 252.

57 ". . . the planning movement in Europe"; ". . . beauty and order": K&M, p. 92.

58 "a gang of scoundrels . . ."; his movements would be closely watched: May 30, 1931, CENTI.

59 Included in a delegation: Ibid.

Chapter 5: Giustizia e Libertà

61 ". . . conflicting interest groups . . .": FRIED, pp. 13–14.

61 ". . . all official institutions . . .": BAR, p. 181.

61 "a dangerous practice . . .": Ibid., p. 223.

61 "as great and powerful . . .": FRIED, pp. 13–14.

62 "a genuine love . . .": BAR, p. 117.

62 ". . . a family affair . . .": HOOP, pp. 176–77.

62 ". . . increasingly vital . . .": Ibid., p. 182.

63 ". . . fraught with threats . . .": To Donald Bachi, April 3, 1935.

64 "[T]he cubic volume . . .": *The Visual Arts: A History,* by Hugh Honour and John Fleming, p. 602.

65 ". . . leaves them out in the rain": SEC, p. 285.

65 ". . . monuments of nothingness . . .": Ibid., pp. 285–86.

66 "The innovation introduced . . .": *Olivetti Builds,* by Patrizia Bonifazio and Paolo Scrivano, pp. 49–50.

68 ". . . veneration and despair . . .": GINZ, p. 77.

69 rescued just in time: Ibid., pp. 84–85 (1935).

69 " '. . . in his overcoat!' ": Ibid., p. 84.

70 "functional city": K&M, p. 92.

73 not so much a master plan: *Olivetti Builds,* p. 105.

74 the slopes of Mont Blanc: CENTI; *Olivetti Builds,* p. 105.

75 anti-Fascist leaflets: CLARK, p. 252.

75 "had considerable appeal": CLARK, p. 252.

75 Another ten thousand fled: *IBM and the Holocaust,* by Edwin Black, pp. 44–45.

77 "The ground was littered . . .": *Christ Stopped at Eboli,* by Carlo Levi, p. 69.

77 ". . . a halo of gold dust . . .": *Un Torinese del Sud: Carlo Levi,* by Gigliola de Donato and Sergio d'Amaro, p. 90.

78 "I'd be dead already . . .": Ibid.

78 "Many business firms . . .": CLARK, p. 238.

79 Little Anna was born: In March of 1938.

79 to give Anna some legal protection: In a letter to the author dated March 28, 2017, Anna Olivetti said she had not been adopted by Adriano Olivetti, but "acknowledged" by him. In any event, she took his name. His motivation "was not to protect a poor little Jewish baby from racial laws, but for a much more subtle and profound reason: to protect Paola unconditionally." He was, she wrote, "acknowledging her basic goodness, in spite of her 'sin' in having brought an extra-marital daughter into the world."

80 "never told me anything . . .": Interview.

80 "Her gestures are full of grace . . .": *Un Torinese del Sud: Carlo Levi,* by Gigliola de Donato and Sergio d'Amaro.

80 ". . . a terrible whore . . .": Ibid., pp. 136–37.

Chapter 6: Terror and Resolve

81 "no one knows . . .": Letter to M. Muntades, October 17, 1939.

84 ". . . The pooled products . . .": *Rumor and Reflection,* by Bernard Berenson, p. 208.

86 ". . . irrelevant frippery . . .": HAR, AM-P, 224, 13.

86 ". . . teeming with intuition . . .": Ibid.

87 ". . . finance and control work . . .": HAR, AM-P, pp. 224, 3.

88 ". . . a year to get him out": Interview with Matteo (Matthew) Olivetti.

88 Ivrea had not: CLARK, p. 289

88 ". . . 'Silly nonsense' . . .": GINZ, p. 134.

89 ". . . what the war was like . . .": Ibid., p. 126.

89 ". . . belly diseases . . .": Letter, July 14, 1941.

89 "courteous manners . . .": *A Spy Among Friends*, by Ben Macintyre, pp. 69–70.

90 "Angleton advised . . .": *A Covert Life*, by Ted Morgan, p. 149.

90 ". . . a bogus source": Ibid., p. 250.

91 ". . . emaciated good looks . . .": *Cloak & Gown* by Robin Winks, p. 170.

91 "appallingly low": Ibid.

91 before the explosion: *James Jesus Angleton, the CIA, & the Craft of Counterintelligence* by Michael Holzman, p. 44.

91 How many potential pyromaniacs miscalculated: Ibid.

91 It was awfully boring: *A Covert Life*, by Ted Morgan, p. 249.

92 "office routine at one end . . .": *James Jesus Angleton*, by Michael Holzman, pp. 44–45.

92 "extremely brilliant but . . .": Ibid., p. 45.

92 "In this game . . .": *Cloak & Gown*, by Robin Winks, p. 545.

93 "a city under siege": *Disciples*, by Douglas Waller, p. 58.

93 "the slow drumming sound": *The Heat of the Day*, by Elizabeth Bowen, pp. 98–100.

94 an alcoholic fervor: *A Spy Among Friends*, by Ben Macintyre, p. 71.

94 ". . . a mentoring relationship . . .": *James Jesus Angleton*, by Michael Holzman, p. 49.

94 ". . . that mythical hell . . .": *Cold Warrior*, by Tom Mangold, p. 40.

94 The rewards were few: Ibid.

94 false information: Ibid.

94 "the incredibly complex problem . . .": *Cloak & Gown*, by Robin Winks, p. 352.

95 ". . . the esoteric secrets . . .": *A Spy Among Friends*, by Ben Macintyre, p. 71.

95 "a single lamp burning . . .": *Cloak & Gown*, by Robin Winks, p. 372.

95 "the finest . . .": Ibid.

95 "From these visitors . . .": Ibid. p. 329.

96 rumored to be working side by side: Ibid., p. 33.

96 "very close" to the ambassador: Ibid., p. 329.

97 "By virtue of . . .": *IBM and the Holocaust*, by Edwin Black, p. 25.

99 the second most important customer: Ibid., p. 111

99 ". . . simple and stunning": Michael Hersh, *Newsweek.*

99 arm was halfway up: *IBM and the Holocaust*, by Edwin Black, pp. 132–33.

99 an autographed copy of a photo of himself: Ibid., p. 70.

Chapter 7: The Brown Affair

100 ". . . a king in exile": GINZ, p. 143.

101 ". . . that fearful, happy look . . .": Ibid., pp. 143–44.

101 Hitler's reckless invasion: CLARK, p. 286ff.

101 "[A]s the war was being lost . . .": CLARK, pp. 295–96.

102 was relatively brief: GINZ, p. 73.

103 the veneer of charm: TAL, p. 16.

103 ". . . She disappeared forever": Ibid.

103 ". . . great personal cruelty . . .": Ibid., p. 9.

103 The only way to get there: The arrival of Dulles in Portugal and his train trip to Switzerland is told on pp. 13–15 of his narrative, *The Secret Surrender.*

105 Dulles took up residence in Bern: TAL, p. 15.

105 "wandered openly . . ."; ". . . espionage bazaar": Ibid., pp. 17–18.

106 The idea began: *The Telegraph,* January 29, 2001.

108 "timorous and remote": CLARK, p. 296.

109 moved to the small town: *The Telegraph,* January 29, 2001.

110 ". . . the ramifications of the war": *Elsa Schiaparelli,* by Meryle Secrest, p. 29.

111 he came away disappointed: *Adriano Olivetti: La Biografia,* by Valerio Ochetto, pp. 110–11.

112 ". . . energetic and gifted . . .": Telegram, May 15, 1943.

112 "We should back him . . .": Telegram, June 19, 1943.

112 ". . . the ruinous results . . .": Letter, June 17, 1943 to V/CD.

114 Operation Husky: Ital Camp., pp. 18–19.

115 ". . . would have been trapped . . .": *The War in Italy 1943–45,* by Richard Lamb, p. 12.

115 "the company of civilized nations": POL, p. 41.

116 "an old owl . . .": *New York Review of Books,* May 23, 2117.

117 ". . . do I know it": David Laskin's article about the house in the via Tasso was published in *The New York Times,* July 12, 2013.

118 the munificent sum: CENTI.

118 The Foreign Office in London turned it down: On July 31, 1943.

118 a complicated game: *Mission Accomplished,* by David Stafford, p. 99.

118 he was arrested: CENTI; letter, Questura di Roma to the Ministry of the Interior, February 28, 1944.

Chapter 8: A Black Crossing

119 reiterated the confidence: Letter in UK archives, October 18, 1943.

120 ". . . she was occupied . . .": CLARK, p. 304.

121 ". . . would see terrible vengeance.": *The Secret War in Italy,* by William Fowler, p. 59.

121 It was time to leave again: Swiss national records, February 9, 1944.

122 ". . . It is naturally desirable . . .": BAR, p. 209.

122 the treatment exhausted him: To "Sister," February 10, 1953.

123 had to be kept warm: To the Family, January 18, 1942.

123 a fine pig: To "Wlady," March 30, 1943.

124 he slept and slept: March 18, 1943.

124 applying for a patent; a day a week in bed: March 30, 1943.

124 take to the barricades: *Camillo Olivetti e il Canavese,* by Dino Alessio Garino, pp. 110–11.

125 He died in a hospital in Biella: On December 4, 1943.

125 where she succumbed: CENTI.

126 "Guiding Jews to safety . . .": BRITISH, p. 109.

126 ". . . floodlights came on . . .": Ibid., p. 110.

126 ". . . a powerful figure . . .": Ibid., p. 109.

128 It took them four days: April 20–24, 1944.

129 ". . . Italian military headquarters": *Bernard Berenson,* by Ernest Samuels, p. 477.

129 Count Ciano had been shot: On January 11, 1944.

129 his atrocious American accent: *Being Bernard Berenson,* by Meryle Secrest, p. 364.

129 finding hiding places: *The Rape of Europa,* by Lynn H. Nicholas, p. 266.

129 "The house is still standing . . .": *War in Val d'Orcia,* by Iris Origo, p. 234.

130 "The city . . . smells . . .": LEW, p. 43.

130 hard for civilians: *War in Val d'Orcia,* by Iris Origo, p. 57.

131 "war in a museum.": *Italian Campaign (World War II),* by Robert Wallace, p. 120.

131 ". . . make little sense . . .": *Rumor and Reflection,* by Bernard Berenson, p. 375.

133 where she sweltered: Ibid., p. 374.

133 ". . . by moonlight yesterday . . .": Ibid., p. 376.

133 "we were kept indoors . . .": Ibid., p. 411.

134 a terrifying experience: Based on interviews with Lidia Olivetti and Albertina Soavi.

134 It would have blown the villa: Ibid.

135 The advice was to evacuate: LEW, pp. 37–38.

Chapter 9: A Wilderness of Mirrors

136 The war was over: CLARK, pp. 314–15.

136 the plant was saved: Storia Olivetti, G. L. Martinoli biography.

137 "[O]nly three workers . . .": Letter, Dino Olivetti, August 15, 1947.

138 "clear, blue, childlike eyes . . .": The comments by Franco Ferrarotti about Adriano Olivetti, his personality, goals, and successes are based on an interview in Rome with the author, April 17, 2016.

138 "mild Mr. Pickwick . . .": C. L. Sulzberger, *The New York Times,* May 16, 1954.

141 In the immediate aftermath: CLARK, p. 317.

141 "Hundreds of thousands . . .": Ibid.

142 "Nothing, absolutely nothing . . .": LEW, p. 29.

142 declared persona non grata: GINZ, p. 81.

142 "The workers' desire . . .": Ibid.

142 he would not know how to do that: CENTI.

143 ". . . shabby and warped . . .": LEW, pp. 30–31.

144 "simply but cleverly . . .": ELECT, p. 26.

146 receive a shrapnel wound: The chronology of the Italian submarine, the *Acciaio,* is based on an account by John Browne published on the Internet under *Submariners' World:* "Regia Marina Italiana," in 2011. Biographical accounts about Ottorino Beltrami in various versions are also on the Internet, as well as in a pseudo-autobiography, *Sul Ponte di Comando,* edited by Alberto de Macchi and Giovanni Maggia, published in 2004. Mr. Beltrami died in 2013.

146 "It is well . . .": Pearson, as quoted in "The Ideological Origins of American Studies," by Michael Holzman.

147 ". . . partner of Allen Dulles.": "Italy's Black Prince," Executive Intelligence Review, by Allen Douglas, p. 8.

147 the operation was shut down: *The Secrets War,* edited by George C. Chalou, p. 224.

148 "As commander of . . .": GAN, p. 64.

148 Angleton received: Ibid.

148 an "Iron Curtain" was descending: *Italy in the Cold War,* by Christopher Duggan and Christopher Wagstaff, p. 37.

149 "The CIA told . . .": WEI, pp. 29–30.

149 vast conventional forces: Ibid., p. 37.

150 "undoubtedly began": GAN, xii.

150 supplying Mussolini's war: Ibid., p. 35.

151 "a wilderness of mirrors . . .": "Italy's Black Prince," Executive Intelligence Review, by Allen Douglas, p. 4.

151 constitution was in place: CLARK, pp. 319–20.

152 an urgent message: On March 5, 1948.

152 ". . . threatened a cataclysm": WEI, p. 31.

152 a "political barricade": Ibid., pp. 31–32.

152 "guerilla movements . . .": Ibid., pp. 32–33.

152 ". . . bizarre recipes . . .": LEW, p. 132.

154 dressed in a negligée: Based on interviews.

Chapter 10: Enigma Variations

155 "wedded to the old conventions . . .": GINZ, p. 167.

156 "My work with Olivetti . . .": adc.global.org/hall-of-fame/giorgio-soavi.

159 "wandering about . . .": This and other comments: interview with Furio Colombo, April 4, 2016.

161 "one of the most instructive . . .": *Lettera 22:Il documentario sul A. Olivetti,* 2009, film.

161 "the leading corporation . . .": Museum of Modern Art bulletin, 1952.

162 sold for $254,500: *New Yorker,* December 7, 2009.

164 " 'The trouble with . . .' ": HAR, AM-P, pp. 228–32.

164 "people would go . . .": Interview with author, September 13, 2016.

167 a coded message: *Life* magazine, March 11, 1958; *New York Times,* March 18, 2001.

169 ". . . we went on skiing": Interview

170 his best white linen suit: Source "Desi"? Au: check

170 ". . . a parachute jump": Interview

170 ". . . it would be hard . . .": HOOP, 165.

171 ". . . 'Because she is' ": To author.

173 ". . . redemption from misery . . .": Letter, undated, c. 1949.

174 even better for calculators: *Olivetti Builds,* by Bonifazio and Scrivano, p. 18.

174 It operated in 117 countries: HAR, 224, p. 1.

174 construction began in 1953: *Olivetti Builds,* by Bonifazio and Scrivano, p. 35.

174 Another nursery school: Ibid., p. 80.

175 "where barnyard animals . . .": *Smithsonian Magazine,* February 2014.

175 cheaply built houses: CENTI.

176 "He chose a poor town . . .": *Fortune* 1959, p. xi.

176 "source of comfort . . .": Speech by Adriano Olivetti on April 23, 1955.

177 ". . . reunites it with nature . . .": *Fortune,* 1959, p. xi.

177 ". . . called a vision . . .": Quoted by Peter Stansky, *Wall Street Journal,* July 29–30, 2017.

Chapter 11: The Experience of a Lifetime

179 ". . . right-wing terrorists": GAN, p. 63.

179 ". . . a clandestine friendship . . .": *The Observer,* January 10, 1993.

180 "the combined Communist . . .": GAN, p. 67.

180 ". . . a subversive fifth column": *Honorable Men: My Life in the CIA,* by William Colby, p. 109.

180 ". . . a lot of hard ball . . .": ORAL, p. 47.

181 "could have run rings . . .": Ibid., p. 11.

181 biggest bunch of roses: Ibid., pp. 13–14.

182 ". . . covert and an overt . . .": Ibid., p. 15.

182 "a very small number": Ibid., p. 16.

182 "No one would dream . . .": Ibid., p. 18.

182 Cuccia took charge: *New York Times* obituary, June 24, 2183.

183 "*the* arc of the Italian network"; ". . . stubborn and vengeful . . .": FRIED, p. 92.

183 ". . . became the tool . . ."; "shifting the contents . . .": Ibid., p. 93.

184 "They will cancel . . . "; favored the interests: Ibid., p. 95.

184 "In the Mafia . . .": Ibid., p. 96.

185 "Not only that, but . . .": This and following comments are from *P101: Quando l'Italia Invento il Personal Computer,* by Pier Giorgio Perotto, in a new translation by Kay Wilson.

187 "unleashed the billions . . ."; "Each day as he went . . .": In *The New York Review of Books,* July 14, 2016.

189 ". . . the best manager . . .": Interview.

190 ". . . the experience of a lifetime": ELEA.

191 ". . . the pet habit . . .": RADICE, p. 57.

191 endless evenings talking; ". . . a mysterious divinity": ELEA.

192 ". . . a fairly nasty . . ."; "couldn't even feel . . .": RADICE, p. 58.

192 "Ettore's blood . . ."; Roberto saved his life: Ibid., p. 62.

194 even more radical: From a conversation with Gastone Garziera.

194 ". . . receive public funding": *Paese Sera,* November 18, 1959.

194 did not like, or approve of: A letter from Allen Dulles to Harold E. Stassen, July 14, 1954.

195 A strong faction . . . objected: POL, p. 255.

196 Comunità won a single seat: CENTI.

196 idiomatic English: "Adriano Olivetti: A 'Socialist' Industrialist in Postwar Italy," by Claudia Nasini, in *Italy and the Bourgeoisie,* ed., Stefania Lucamente, pp. 96–97.

196 the kind of amused tolerance: Allen Dulles, letter to Harold E. Stassen, July 14, 1954.
197 "Grazia had a problem . . .": This and other comments come from an interview with Domenico de' Liguori Carino in Rome.
198 ". . . we just ate the fender": This and other comments come from two interviews with Milton Gendel in Rome, April 4, 2016.
198 "I have no talent . . .": *Vanity Fair,* November 2011.

Chapter 12: High Stakes

200 ". . . impelled to discover": LA, n.d.
200 ". . . restraining Adriano": *Fortune,* September 1960, p. 241.
201 "like Mecca to the Arabs . . .": Interview with Milton Gendel.
201 ". . . by then it's too late": Mediapost.com, June 20, 2013.
201 "a mood of deep depression": *Fortune,* September 1960, p. 242.
201 ". . . We are losing money . . .": Ibid.
202 "worth keeping alive": Ibid.
202 the key to opening up: HAR, pp. 225–21.
202 Despite Underwood's loss: *Fortune,* September 1960, p. 242.
203 family members changed their minds: Ibid., p. 245.
206 ". . . rather grand": Interview with Milton Gendel.
206 "insistent voices and controversy . . .": PER, no pagination.
208 Mario Tchou was dead: Based on interviews.
210 ". . . impossible to determine . . .": *La Stampa,* November 10, 1961.
211 ". . . era of joyful work . . .": ELECT, p. 10.
211 "What are we going to do? . . .": Anecdote by David Olivetti.
211 "a congenial machine . . .": PER.
213 "This left the structural setup . . .": From the Kay Wilson translation, no pagination.
214 ". . . any conflicts and problems": PER.
215 "With unimaginable suddenness . . .": Letter to author, July 26, 2016.
215 ". . . this rumor circulated": *L'Unità,* February 21, 1964.
215 ". . . just a business": *Financial Times,* May 29, 1964.
216 "they did not care a fig": PER translation.
216 It must be eradicated: Speech, April 30, 1964.
216 "They lied to us": ELEA.
217 ". . . the mechanical era is over": PER, no pagination.
218 "the prototype of the Programma 101 had been stolen": The event was related by Roberto Battagazzorre, the son of the Ivrea police chief, on September 26, 2016.
224 "first lonely brick" PER.
225 "letting itself be dragged along . . .": PER, pp. 30–31.
226 The IBM 610: arstechnica.com, June 21, 2011.

Chapter 13: The Curious Case of the Second Death

230 "keep his mouth shut . . .": BAR, p. 157.

230 ". . . facts prove illusory . . .": HOOP, p. 54.

230 this doctor could not be sure: *New York Times,* September 17, 1975.

233 "meant to look like death . . .": *The Annals of Unsolved Crime,* by Edward Jay Epstein, p. 12.

233 "omicidi eccellenti": *Savage Continent,* by Keith Lowe, p. 285.

233 "the arranged accident": *The Annals of Unsolved Crime,* by Edward Jay Epstein, p. 12.

234 "crimes of state . . .": Ibid., p. 15.

235 "As much as anyone . . .": *New York Review of Books,* July 14, 2016.

236 ". . . foreign rivals . . .": March 17, 2018, p. 11.

236 "the country's biggest fear . . .": "History of Computers in the Military," unc.edu.

237 she said, yes he did: To author, in interview.

238 ". . . allied collusion . . ."; ". . . a terrible situation . . .": *Daily Telegraph,* December 20, 2008.

240 some kind of accident: Matteo Olivetti.

240 ". . . a general feeling": Interview.

241 division was a "mole": wordpress.com, July 28, 2011.

241 in its probing report: Mario Caglieris and Enrico Cuccia have since died.

241 "much worse than it actually was?": Interview [with De Benedetti]

242 "The latter is . . .": HOOP, p. 223–24.

242 ". . . the Agnelli-led clique . . .": FRIED, p. 113.

242 "you didn't need . . .": Interview with David Olivetti.

243 "conservative elements . . .": POL, pp. 264–65.

243 ". . . the Moscow banana peel . . .": TAL, p. 465.

243 "a center-left government . . .": POL, p. 259.

244 ". . . would introduce reforms . . .": "Italy's Black Prince," *Executive Intelligence Review,* by Allen Douglas, February 4, 2005.

244 a resurgence in their value: Interviews.

244 "was earning a profit . . .": HAR, AM-P, p. 225.

244 a little more than six millions: Ibid.

245 " 'You thought they were different?' "; "a big missed opportunity": ELEA.

245 ". . . The psychological aspects . . ."; "for political reasons": HAR, AM-P, p. 233.

246 "Then they forced . . .": Interview with author in Rome, April 7, 2016.

248 Olivetti's "Mata Hari": From articles in *La Repubblica* of July 14–18, 1990, July 22, 1990, October 5, 1991, and August 8, 2005.

249 "Nothing happened . . .": To author.

249 "have been used . . .": *Science* magazine, August 14, 1964.

249 "He not only offered . . ." Edward Jay Epstein, p. 109.

250 he believed French Intelligence responsible, Edward Jay Epstein, p. 112.

251 . . . built the teletype machines: PER.

253 ". . . he really got mad . . .": Interview with David Olivetti.

254 a lengthy description of his hiring: On May 5, 2016, in Turin.

255 "Productivity and morale . . ."; the second largest: *New York Times,* February 19, 1984.

255 "He was treated . . .": Interview with David Olivetti.

255 "Roberto Olivetti was weak . . .": Ibid.

255 "brimming with . . . vitality": *Scottsass, Olivetti, Synthesis:* ed. by Morteo, Saibene, Meneguzzo, and Carboni, p. 119.

256 ". . . he was at least . . .": Ibid.

256 if he needed a chair: Desire? Au: check note

256 light, but no view: SEC, p. 444.

260 ". . . morally rigorous . . .": Ibid. p. 149.

263 ". . . a good sign": ELEA.

263 "The factory not only": To author, email, July 24, 2018.

263 ". . . the peace of mind . . .": *Ruskin Today,* by Kenneth Clark, pp. 264–65.

263 ". . . the whole style of attack . . .": Ibid.

264 "the central part of him": source?

265 As Morris wrote: Quotes from William Morris and quotes about his character and ideals come from the biography of Morris: *William Morris: A Life for Our Time,* by Fiona MacCarthy.

Bibliography

Aarons, Mark, and John Loftus. *Unholy Trinity: The Vatican, the Nazis, and the Swiss Banks*. St. Martin's Griffin, 1998.

Agee, Philip, and Louis Wolf, eds. *Dirty Work: The CIA in Western Europe*. Zed Press, 1978.

Amoroso, Bruno, and Nico Perrone. *Capitalismo Predatore*. Castelvecchi RX, 2014.

Barzini, Luigi. *The Italians*. Atheneum, 1964.

Bashe, Charles J., Lyle R. Johnson, John H. Palmer, and Emerson W. Pugh. *IBM's Early Computers*. MIT Press, 1986.

Beltrami, Ottorino. *Sul Ponte di Comando: dalla marina militare alla Olivetti*. Mursia, 2004.

Bertoni, Ugo. *Capitalisti d'Italia*. Boroli Editore, 2003.

Bilo, Federico, and Ettore Vadini. *Matera e Adriano Olivetti*. Edizioni di Comunità, 2013.

Black, Edwin. *IBM and the Holocaust*. Dialog Press, 2011.

Blum, William. *Killing Hope: U.S. Military and CIA Interventions Since World War II*. Common Courage Press, 1995.

Bonifazio, Patrizia, and Paolo Scrivano. *Olivetti Builds*. Skira, 2001.

Bricco, Paolo. *L'Olivetti dell'Ingegnere*. Il Mulino, 2014.

Cadeddu, Davide. *Reimagining Democracy*. Springer, 2012.

Caizzi, Bruno. *Gli Olivetti*. Unione Tipografico—Editrice Torinese, 1962.

Ceruzzi, Paul E. *Computing: A Concise History*. MIT Press, 2012.

Chalou, George C., ed. *The Secrets War: The Office of Strategic Services in World War II*. National Archives and Records Administration, 1992.

Clark, Jennifer. *Mondo Agnelli: Fiat, Chrysler, and the Power of a Dynasty*. John Wiley & Sons, 2012.

Clark, Martin. *Modern Italy: 1871–1995*. Longman, 1998.

Clarridge, Duane R. *A Spy for All Seasons*. Scribner, 1997.

Colby, William. *Honorable Men: My Life in the CIA*. Simon & Schuster, 1978.

Corvo, Max. *OSS Italy, 1942–1945*. Enigma Books, 1990.

Cosenza, Luigi. *The Olivetti Factory in Pozzuoli*. CLEAN Edizioni, 2006.

Davis, John A., and Paul Ginsborg. *Society and Politics in the Age of the Risorgimento*. Cambridge University Press, 1991.

De Donato and Sergio D'Amaro. *Un Torinese del Sud: Carlo Levi.* Baldini & Castoldi, 2001.

De Wit, Wim, ed. *Design for the Corporate World, 1950–1975.* Lund Humphries, 2017.

Duggan, Christopher, and Christopher Wagstaff. *Italy in the Cold War: Politics, Culture and Society.* Berg, 1995.

Dulles, Allen. *The Secret Surrender.* Harper & Row, 1966.

Epstein, Edward Jay, *The Annals of Unsolved Crime.* Melville Press, 2012.

———. *James Jesus Angleton: Was He Right?* FastTrack Press, 2013.

Feltrinelli, Carlo. *Senior Service: A Study of Riches, Revolution and Violent Death.* Granta Books, 1999.

Ferrarotti, Franco. *I Miei Anni con Adriano Olivetti.* Solfanelli, 2016.

Friedman, Alan. *Agnelli and the Network of Italian Power.* Mandarin, 1988.

Ganser, Daniele. *NATO's Secret Armies: Operation Gladio and Terrorism in Western Europe.* Frank Cass, 2005.

Garino, Dino Alessio. *Camillo Olivetti e il Canavese tra Ottocento e Novecento.* Le Chateau Edizioni, 2004.

Gerbi, Sandro. *Giovanni Enriques dalla Olivetti alla Zanichelli.* Editore Ulrico Hoepli, 2013.

Ghiringhelli, Peter. *A British Boy in Fascist Italy.* The History Press, 2010.

Ginsborg, Paul. *A History of Contemporary Italy: Society and Politics 1943–1988.* Palgrave Macmillan, 2003.

Ginzburg, Natalia. *Family Sayings.* Arcade Publishing, 1963.

———. "Family Lexicon." *New York Review of Books,* 2017.

———. *Voices in the Evening.* Arcade Publishing, 1961.

Giusti, Antonio. *La Casa del Forte dei Marmi.* Le Lettere, 2008.

Granata, Cora, and Cheryl A. Koos. *Modern Europe, 1750 to the Present.* Rowman & Littlefield, 2008.

Holmes, Douglas R. *Cultural Disenchantments: Worker Peasantries in Northeast Italy.* Princeton University Press, 1989.

Holt, Thaddeus. *The Deceivers: Allied Military Deception in the Second World War.* Skyhorse Publishing, 2007.

Holzman, Michael. *James Jesus Angleton, the CIA, & the Craft of Counterintelligence.* University of Massachusetts Press, 2008.

Hooper, John. *The Italians.* Penguin Books, 2015.

Johnson, Loch K. *America's Secret Power: The CIA in a Democratic Society.* Oxford University Press, 1989.

Kalanithi, Paul. *When Breath Becomes Air.* Random House, 2016.

Kargon, Robert K., and Arthur P. Molella. *Invented Edens: Techno-Cities of the Twentieth Century.* MIT Press, 2008.

Kempe, Frederick. *Berlin 1961: Kennedy, Khrushchev, and the Most Dangerous Place on Earth.* G. P. Putnam's Sons, 2011.

Kicherer, Sibylle. *Olivetti: A Study of the Corporate Management of Design.* Trefoil Publications, 1990.

Lewis, Norman. *Naples '44.* Carroll & Graf, 2005.

Lucamente, Stefania, ed. *Italy and the Bourgeoisie: The Re-Thinking of a Class.* Farleigh Dickinson University Press, 2009.

MacCarthy, Fiona. *William Morris: A Life for Our Time.* Alfred A. Knopf, 1994.

Macintyre, Ben. *A Spy Among Friends: Kim Philby and the Great Betrayal.* Broadway Books, 2014.

Mangold, Tom. *Cold Warrior: James Jesus Angleton: The CIA's Master Spy Hunter.* Simon & Schuster, 1991.

Martin, David C. *Wilderness of Mirrors: How the Byzantine Intrigues of the Secret War Between the CIA and the KGB Seduced and Devoured Key Agents James Jesus Angleton and William King Harvey.* Harper & Row, 1980.

McNamara, Joel. *Secrets of Computer Espionage: Tactics and Countermeasures.* Wiley Publishing, 2003.

Migone, Gian Giacomo. *The United States and Fascist Italy: The Rise of American Finance in Europe.* Cambridge University Press, 2015.

Minardi, Mario. *Il Canavese Ieri e Oggi.* Ilte Torino, 1960.

Mistry, Kaeten. *The United States, Italy and the Origins of Cold War.* Cambridge University Press, 2014.

Morgan, Ted. *A Covert Life: Jay Lovestone: Communist, Anti-Communist, and Spymaster.* Random House, 1999.

Morteo, Enrico. *Mario Bellini: Furniture, Machines and Objects.* Phaidon, 2015.

Nasheri, Hedieh. *Economic Espionage and Industrial Spying.* Cambridge University Press, 2005.

Newark, Tim. *Mafia Allies: The True Story of America's Secret Alliance with the Mob in World War II.* Zenith Press, 2007.

Ochetto, Valerio. *Adriano Olivetti: La Biografia.* Edizioni di Comunità, 1985.

Olivetti, Adriano. *Dall'America: lettere ai familiari (1925–1926).* Edizioni di Comunità, 2016.

———. *Ai Lavoratori.* Edizioni di Comunità, 2013.

Olivetti, Camillo. *Lettere Americane.* Fondazione Adriano Olivetti, 1999.

———. *Camillo e Luisa, 1899–1943.* Fondazione Adriano Olivetti, 1999.

Olivetti, Desire, ed. *Roberto Olivetti.* Edizioni di Comunità, 2003.

Olivetti, Massimo. *Per Viver Meglio.* Stab.Tip. de "Il Giornale d'Italia," 1949.

O'Reilly, Bill. *Killing Patton: The Strange Death of World War II's Most Audacious General.* Henry Holt, 2014.

Origo, Iris. *War in Val d'Orcia: An Italian War Diary, 1943–1944.* David R. Godine, 2000.

Peroni, Marco. *Ivrea: Guida alla Città di Adriano Olivetti.* Edizioni di Comunità, 2016.

Perotto, Pier Giorgio. *P101: Quando l'Italia Invento il Personal Computer.* Edizioni di Comunità, 1995.

Perry, William J. *My Journey to the Nuclear Brink.* Stanford University Press, 2005.

Pisani, Sallie. *The CIA and the Marshall Plan.* University Press of Kansas, 1991.

Powers, Thomas. *The Man Who Kept the Secrets: Richard Helms and the CIA.* Alfred A. Knopf, 1979.

Prados, John. *Safe for Democracy: The Secret Wars of the CIA.* Ivan R. Dee, 2006.

Radice, Barbara. *Ettore Sottsass: A Critical Biography.* Rizzoli, 1993.

Ripa di Meana, Carlo. *Le Bufale.* Maretti Editore, 2014.

Saibene, Meneguzzo and Carbene. *Sottsass, Olivetti, Synthesis.* Edizioni di Comunità, 2016.

Stafford, David. *Mission Accomplished: SOE and Italy, 1943–1945*. The Bodley Head, 2011.

Stockton, Bayard. *Flawed Patriot: The Rise and Fall of CIA Legend Bill Harvey*. Potomac Books, 2006.

Talbot, David. *The Devil's Chessboard: Allen Dulles, the CIA, and the Rise of America's Secret Government*. HarperCollins, 2015.

Tompkins, Peter. *A Spy in Rome*. Avon Books, 1962.

Wallace, Robert, and H. Keith Melton. *Spycraft: The Secret History of the CIA's Spytechs, from Communism to Al-Qaeda*. Plume, Penguin Books, 2009.

Waller, Douglas. *Disciples: The World War II Missions of the CIA Directors Who Fought for Wild Bill Donovan*. Simon & Schuster, 2015.

Watson, Thomas J. Jr., and Peter Petre. *Father, Son & Co., My Life at IBM and Beyond*. Bantam Books, 1990.

Weinberger, Sharon. *The Imagineers of War: The Untold Story of DARPA, the Pentagon Agency That Changed the World*. Alfred A. Knopf, 2017.

Weiner, Tim. *Legacy of Ashes: The History of the CIA*. Anchor Books, 2008.

Willan, Philip. *Puppet Masters: The Political Use of Terrorism in Italy*. Constable, 1991.

Williams, Paul L. *Operation Gladio: The Unholy Alliance Between the Vatican, the CIA and the Mafia*. Prometheus Books, 2015.

Winks, Robin W. *Cloak & Gown: Scholars in the Secret War, 1939–1961*. Yale University Press, 1987.

Index

Page numbers in *italics* refer to photographs.

Acciaio, 144–46
Action Party, 107, 110, 141
Acton, Harold, 133
Adler, Alfred, 55
Adrema Werke, 202
Agnelli, Fiat and the Network of Italian Power (Friedman), 183–84
Agnelli, Gianni, 43, 142, 181, *181, 183,* 184, 195, *247,* 254
Agnelli and the Network of Italian Power (Friedman), 61
Agnelli family, 61, 62, 184, 242, 254
Air Force, U.S., 187, 224, 235
Albert I of Belgium, 106
Alcoa, 98
al-Mabhouh, Mahmoud, 233
Alps, 5, *10,* 71, 73, *73,* 127, 218, 262
Amoroso, Bruno, 240
Andreotti, Giulio, 150
Angleton, James Hugh, 89, *90,* 95–96, 146–47
Angleton, James Jesus "Jim," 89–96, *92,* 146–48, 150, 243
Angleton, Carmen Mercedes Moreno, 89
Annals of Unsolved Crime, The (Epstein), 233–34
Antoni, Carlo, 106
Aosta, 73, 74

Apollo 11, 223–24
Apple, 226
Appointment with Death (Christie), 228
architecture, 63–67, 70–72, 86, 159
Architecture of Humanism, The (Scott), 84
Armory Show, 63
art, 63
Art Deco, 50, 73
Ashbee, C. R., 37–38
atomic bomb, *see* nuclear weapons
Attlee, Clement, 138

Bacciagaluppi, Giuseppe, 126
Badoglio, Pietro, 108, *108,* 115, 117, 118, 119, 124, 141, 147
Banca d'Italia, 245
Banfi, Gian Luigi, 70
Barzini, Luigi, 61, 230
Basilicata (Lucania), 76–77
Battegazzorre, Giuseppe, *183,* 218, *220,* 239, 247
Battegazzorre, Roberto, 218, 239, 246
Battle of the Oranges, 5–11, *6, 7, 8,* 13, 15
Baudoun, Charles, 55
Bauer, Riccardo, 68
Bauhaus, 64–67, 86, 256
Bazata, Douglas, 238

BBPR, 70, 74
Belgiojoso, Ludovico Barbianodi, 70
Belgium, 63
Bellin, Robinson O., 147
Bellini, Mario, 217–18
Beltrami, Ottorino, 14, 143–46, *143,*
    150, 254–55
Berenson, Bernard, 83, 84–85, 128–29,
    131, 133
Berenson, Mary, 133
Beria, Ricardo, 168, 205
Beria, Vittoria, 168, 171, 197–98, 205
Berlin, 54
Berlin Wall, 208, 234
Berlusconi, Silvio, 62
Bernardo, St., 31
Black, Edwin, 97, 99
Black Nobility, 147
Black Prince (Junio Valerio Borghese),
    147–48, 150, 179, 244
Blum, William, 180
Bonifazio, Patrizia, 29, 73–74
Bonomi, Ivanhoe, 107, 141
Borghese, Junio Valerio (the Black
    Prince), 147–48, 150, 179, 244
Borielli, Lucius, 189–90
Bowen, Elizabeth, 93
Bretti, Franco, 212–13
Breuil valley, 74
Brosio, Manlio, 107
Brown, Jerry, 187–88, 235
Brussels International Fair, 47
Bucci-Casari, Elisa, 171, 259
Buck, Dudley, 235
Burnham and Root, 20
Burzio, Domenico, *23,* 24, 28, 51, 52
Business Week, 222

Cagliari, 145–46
Caglieris, Mario, 241
Calogero, Gino, 144
Calogero, Giuseppe, 186, 211
Canavese, 19–20, 200
capitalism, 61, 183, 242, 243, 263, 265
Cappellaro, Natalino, 162, 217, 221
Capponi, Iginio, 258

Carandini, Nicolò, 106–7
Caronia, RMS, 50
Caruso, Bruno, 158–59, *260*
Casey, Bill, 93
Castro, Fidel, 180
Catholic Church, 25
Caviglia, Enrico, 107
Cavour, Camillo Benso, Count, 17
Celani, Claudio, 243–44
Census Bureau, U.S., 97
Ceruzzi, Paul E., 226
CGS, 24, 27, 28
Chiang Kai-shek, 185
Chicago World's Fair, 20–21
China, 204, 206–8, 236, 237, 240,
    246
Christian Democrats, 151–52, 243
Christie, Agatha, 228, 232
Christ Stopped at Eboli (Levi), 76–77,
    174
Church, Frank, 232, *234*
Church Committee, 180, 232, *234*
Churchill, Winston, 113, 115, 138, 148,
    149, 157
CIA (Central Intelligence Agency),
    89–93, 103, 144, 146, 147, 149,
    152, 179, 180, 194–96, 232, 237,
    243, 254
  poison dart guns of, 232, *234*
CIA and the Marshall Plan, The (Pisani),
    146
Ciano, Gian Galeazzo, 84, 129
cities:
  functional, 70
  garden, 58
Clark, Kenneth, 263
Clark, Martin, 22, 56–57, 75, 78, 101,
    108, 120, 141–42
Clay, Lucius D., 152
Cloak & Gown (Winks), 91
Cocteau, Jean, 66
Cola di Rienzo, 62
Colby, William, 144, 180, 232
Cold War, 148–50, 178–79, 181, 234,
    236, 240, 243, 260
Colombo, Furio, 159–61

Columbia University, 225
Columbus, Christopher, 20
Comitato Nazionale per l'Energia
  Nucleare (CNEN), 251
Committee for National Liberation
  (CLN), 126
Committee of National Liberation for
  Upper Italy (CLNAI), 136, 141, 142
Communism, 39, 44, 56, 75, 88, 91,
  94, 110, 137, 139, 142, 148–52,
  178–82, 194–95, 204, 237, 243
Comunità, 110–12, 139–40, 151, 152,
  194–96, 200, *202,* 237
Comunità, 156–57
Computer History Museum, 225
computers, 28, 186, 206, 217, 224, 225,
  247
  cryotrons and, 235–36
  ENIAC, 187
  General Electric, 215–16
  Hewlett-Packard, 224
  IBM, 11, 186, 226, 235, 246; *see also*
    IBM
  Kenbak-1, 225–26
  Olivetti, *see* Olivetti computers
  punch cards for, 97–98, 186, 212, 217
  transistors in, 11, 190, 193–94
Computing: A Concise History (Ceruzzi),
  226
Confindustria, 243
Contadini, Pietro, 191, 238
Cosenza, Luigi, 176
Croce, Benedetto, 106, 107
Cryotron Files, The (Dey and Buck), 236
cryotrons, 235–36
Cuban Missile Crisis, 187, 188, 234
Cuccia, Enrico, 182–84, *182,* 241, 242,
  244
Cutting, Sybil, 84, 129
cyber security, 236–37

D'Annunzio, Gabriele, 120, *120*
Davis, Norman H., *114*
Davis, Paul, 156
Davis, Richard Harding, 109–10
D-Day, 91, 92, 115

de Benedetti, Carlo (friend of Paola),
  45–46
De Benedetti, Carlo (Olivetti CEO),
  234, 241, 246, *247,* 255, 259
de Charrière, Isabelle, 84
De Gasperi, Alcide, 151, 175
Dehomag, 98
de' Liguori Carino, Beniamino, 172,
  263
de' Liguori Carino, Domenico, 196–97,
  199
De Mauro, Mauro, 250
Depression, Great, 56
De Sandre, Giovanni, *212*
Despard, Caroline Scott, 3
de Staël, Madame, 85
Devil's Chessboard, The (Talbot), 103
Diaghilev, Sergei, 156, 164
Dimitrev, Victor, 248
Divisumma, 162, 217
Donovan, William "Wild Bill," 92, 238
Douglas, Allen, 147, 151
Duchamp, Marcel, 63
Dulles, Allen, 102–6, *102,* 118, 125, 147,
  188, 243
  Adriano Olivetti and, 102, 110–11,
    194–96, 237
Dunn, James, 152

Ecclesia, Edward, 213
Economist, 236
Eden, Anthony, 113–14, *114,* 115
Edison General Electric, 21
Einaudi, Luigi, 107
Eisenhower, Dwight D., 92, 195
Elea computer, 186, 189–91, 193–94,
  204, 206, 209, 238
electricity, 20, 21, 23
  Camillo Olivetti's device for
    measuring, 23–24, 27
electronics, 214, 221, 225, 251
  at Olivetti, 184–86, 188–92, 204,
    206, 211, 215–16, 220, 229, 237,
    240–41, 245, 246, 254, 260
  see also computers
Elisabeth of Bavaria, 106

Elizabeth II, Queen, 198
ENIAC, 187
Enriques, Giovanni, 136, 137
Ente Nazionale Idrocarburi (ENI), 249
Epstein, Edward Jay, 233–34, 250
Executive Intelligence Review, 147

Fairchild Semiconductor Company,
    224, 225
Family Sayings (Ginzburg), 33, 40
Fanfani, Amintore, 195
Fascism, 44, 48, 56–59, 63, 68–69, 75,
    88, 98, 101–2, 107, 111, 115, 116,
    121–22, 124, 125, 136, 137, 141,
    142, 146–48, 150, 179, 182
FBI (Federal Bureau of Investigation), 90
Feltrinelli, Giangiacomo, 169
Feltrinelli, Inge, 259–60
Fermi, Enrico, 78, 185
Ferraris, Galileo, 20, 21, 23
Ferrarotti, Franco, 137–41, 167, 197, 199
Ferrero, 62
Fiat, 18, 42–43, 61, 141, 142, 144–45,
    150, 159, 180–81, *181, 183, 183,*
    184, 215, 216, 240–42, 244, 254
Fichera, Massimo, 263
Fiesole, 82–84, *83,* 131–34
Figini, Luigi, 65, 66, 70, 71, 74, 174
Filipassi, Franco, 190
Fina, Thomas W., 181–82
Financial Times, 215
Fininvest, 62
Fleming, John, 64
Florence, 130–31, 133, 134
Foà, Luciano, 87
Foà, Vittorio, 68
Folon, Jean-Michel, 156
Fondazione Adriano Olivetti, 72, 172,
    258
Ford Foundation, 195, 198
Ford Motor Company, 52, 98
Forrestal, James V., 152
Fortini, Franco, 159
Fortune, 11, 174, 176, 178, 200, 201,
    202, 222
forza di un sogno, La, 37, 230–31

Fowler, William, 121
France, 101, 102, 104, 105
Francis of Assisi, St., 25
Franco, Francisco, 63
French Revolution, 5, 38
Freud, Sigmund, 55, 87, 259
Friedman, Alan, 60–61, 183–84, 242,
    244
Friedman, Martin, 186
Frinzi, Francesco, 209–11, 240
Fua, Giorgio, 87
Futurism, 63, 221

Gabetti, Gianluigi, *143, 207,* 254, 255,
    257
Gaiani, Antonio, 119
Galardi, Alberto, 215
Galardi, Maria Luisa "Mimmina," 36,
    37, 123, 153, 215
Galassi, Ugo, *143,* 163–64, 201
Galleria della Confederazione Nationale
    Artisti Professionisti, 74
Galletti, Germana, 172
Galletti, Paolo, 172
Galletti, Remo, 186
Gandhi, Mahatma, 263
Ganser, Daniele, 147–48, 178–79,
    235
Garino, Dino Alessio, 125
Garziera, Gastone, *212,* 216,
    248–49
Gendel, Milton, 168, 197–98, 201,
    205–6, *205*
Gendel, Natalia, *205*
Gendel, Sebastiano, *205*
General Electric, 21, 215–16, 235, 241,
    251, 254
General Fascist Confederation of
    Industrialists, 59
General Motors (GM), 98
Germany, 148, 179
    Berlin Wall in, 208, 234
Germany, Nazi, *see* Nazi Germany
Gerrarotti, Franco, 13–14
Ghiringhelli, Peter, 125–26
Ginsborg, Paul, 142, 195, 243

Ginzburg, Leone, 68, 75, 76, 79, 100, 116–17
  death of, 117
  marriage of, 68, 79
Ginzburg, Natalia Levi, 33, 35, 36, 40, 45, 49, 68, 69, 75, 79, 85, 88–89, 100–101, 116, 155
  marriage of, 68, 79
Giudici, Giovanni, 159
Giustizia e Libertà, 68–69, 75–77, 107, 261
Gladio, 150, 179, 246, 254
Glaser, Milton, 156
Gorbachev, Mikhail, 148
Gourevitch, Vita, 77
Grass, Günter, 162
Great Depression, 56
Gropius, Walter, 63, 64
Gruppo 7, 65
Guarracino, Ottavio, 211
Guicciardini, Francesco, 253
Guttuso, Renato, *78*
Gypsies, 98–99

Hamas, 233
Harvard University, 43, 48, 163, 166, 185, 241, 244, 245
Hearst, William Randolph, 99
Helms, Richard, 92
Hemingway, Ernest, 87
Herskovitz, Agata, 126
Hewlett-Packard (HP), 224
Hindenburg, Paul von, 63
Hitchcock, Alfred, 232
Hitler, Adolf, 63, 98, 99, 131, 136, 147
  Mussolini and, 78, 101, 120
  Watson and, 99
Hodgson, Richard, 224–25, *226*
Hollerith, Herman, 97–98
Holzman, Michael, 89, 92
Honour, Hugh, 64
Hooper, John, 25, 37, 62, 170–71, 230, 241–42
Hotchkiss, Gord, 201
Hubbard, Elbert, 38

Iacocca, Lee, 181
IBM (International Business Machines), 11, 96–99, 186, 187, 188, 190, 194, 202–3, 221, 226, 235, 236, 241, 246, 255
  Nazi Germany and, 98–99, 116
  SAGE and, 187
IBM and the Holocaust (Black), 97, 99
ICBMs, 208, 234
Ichino, Anna Maria, 77
IFIL, 254
I. G. Farben, 98
Industrial Revolution, 38
International Style, 64
Invented Edens (Kargon and Molella), 70
Ippolito, Felice, 250–51
IRI, 182, 245
Isola, Aimaro Oreglia d', 257
Istituto Mobiliare Italiano (IMI), 245
Italia Libera, 100
Italians, The (Barzini), 61, 230
Italians, The (Hooper), 25, 37, 170–71, 240–41
Italians and the Holocaust, The (Zuccotti), 116
Italy:
  elections in, 151–52, 178, 180, 181, 195, 243
  family businesses in, 62
  in fourteenth century, 62
  garden cities in, 58
  German occupation of, 115–17, 121, 124, 126, 129–35, 136–37
  interest groups in, 61
  "Mafia" concept in, 60–61
  Mussolini in, *see* Mussolini, Benito
  in postwar period, 141–42, 151
  smuggling between Switzerland and, 218
  Soviet Union and, 243
  unification of, 18, 61
  U.S. and, 151–52, 178, 180, 181, 243, 243, 248–51
  in World War I, 42–43, 44
  in World War II, *see* World War II

Ivrea, 5–6, 52, 67, 88, 102, 121, 124, 136, 137, 139, 142, 196
architecture in, 71–72
Battle of the Oranges in, 5–11, *6, 7, 8,* 13, 15
cemetery in, 262
memorial to Adriano Olivetti in, 262
memorial to Camillo Olivetti in, 17
Olivetti company and, 9, 11
reorganization plans for, 71, 73
World Heritage Site of, 263

James, Henry, 264
James Jesus Angleton, the CIA & the Craft of Counterintelligence (Holzman), 92
Japan, 237, 255
Jarach, Bruno, 252
Jervis, Guglielmo "Willy," 261–62
Jewish mothers, 170–71
Jews, 63, 75, 78, 79, 84, 98–99, 116, 125, 126, 137
Jobs, Steve, 226
Johnson, Philip, 64–65
Jung, Carl, 55, 87, 259

Kahn, Louis, 256
Kargon, Robert, 70
Keats, John, 17
Kenbak-1, 225–26
Kennan, George, 149
Kennedy, John F., 188, 208–9, 233, 243, 249
Kesselring, Albert, 130–31
Keynes, John Maynard, 87
Khrushchev, Nikita, 178
Kierkegaard, Søren, 87
Killing Hope (Blum), 180
Kim Koo, 180
Korean War, 187

La Malfa, Ugo, 107
La Martella, 175
Lamb, Richard, 115
Lanfranco, Fausto, 240
Legacy of Ashes (Weiner), 149

La Piazzola, 82–84, *83,* 129, 133, 134
La Serra complex, 72, 258, *264*
Laskin, David, 116
Lattes, Franco, 159
Le Corbusier, 63–67, *65,* 70, 73, 159
Levi, Alberto, 40
Levi, Carlo, 75–80, *78,* 107, 116, 174–75
child born to Paola and, 79–80
at Lucania, 76–77
paintings of, 76, *76,* 77, 79
Paola's relationship with, 77–80
writings of, 76
Levi, Gino (Gino Martinoli), 40, 56, 62, 75–76, 136, 137, 168, 255
Levi, Giuseppe, 40–42, 45, 46, 54, 55–56, 68, 75, 79, 88, 132, 134
Levi, Lidia Tanzi, 40, 45, 46, 88, 132
Levi, Mario, 40–42, 68–69, 75, 77
Levi, Natalia, *see* Ginzburg, Natalia Levi
Levi, Paola, *see* Olivetti, Paola Levi
Levi family, 40, 49, 50, 53, 68, 75
Lewis, Norman, 130, 134–35, 142, 143, 152–53
Libya, 101, 144
Life, 166, 252
Lindbergh, Charles, 99
Linder, Erich, 87
Lizier, Carlo, 62, 124, 153
Lizier, Laura "Lalla" Olivetti, 30, 31, *34, 36,* 39, *41,* 62, 78, 153
Lizier, Mimmina, *see* Galardi, Maria Luisa "Mimmina"
Lockheed Corporation, 235
Logos 27 calculator, 220, 221, 225
London, 53, 54, 93
Lucania, 76–77
Luce, Clare Boothe, 178, *179,* 181, 243
Luce, Henry, 178
Luciano, Charles "Lucky," 179
Luetscher, John, 192
Lussu, Emilio, 107

MacCarthy, Fiona, 265
Machiavelli, Niccolò, 61, 102, 144, 206
Macintyre, Ben, 89
Mackail, J. W., 264

Mafia, 179, 184, 250, 254
"Mafia" concept, 60–61
Mainardis, Pietro, 258
Mangold, Tom, 94
Mantegazza, Paolo, 39
Man with the Poison Gun, The (Plokhy), 232
Mao Zedong, 204
Marchesini, Maria, 77
Mariano, Nicky, 128–29
Marie José of Belgium, 106, *106,* 107, 109
Mario, Pier, 168
Mariotti, Roberto, 247–48
Marotta, Domenico, 250
Marrighini, Luca, 258
Marshall Plan, 146, 150, 152, 180, 198, 241, 254
Martinoli, Gino, *see* Levi, Gino
Marzotto, 194
Matera, 174–75, 176
Mattei, Enrico, 233, 249–50, *250,* 251
Matteotti, Giacomo, 48–49, *48*
Matterhorn, 74
Mazzini, Giuseppe, 60
McCarthy, Cormac, 162
McCarthy, Joseph, 180
McElheny, Victor K., 249
Medici, Cosimo de', 83, 131
Medici, Giovanni de', 83
Medici, Lorenzo de', 83
Mediobanca, 182–84, 242, 244–45
Metraux, Guy, 15
Meyer, Adolf, 63
Meyerson, Bernie, 226
Mies Van der Rohe, Ludwig, 63, 65
Milan, 61, 67, 87, 88, 101, 121, 136
fair of 1958 in, 189
missiles, 187, 188, 234
ICBMs, 208, 234
MIT (Massachusetts Institute of Technology), 235, 236
Modernism, 72
Modern Italy (Clark), 75
Mole City (Talponia), 257–58, *257*
Molella, Arthur, 70

Montanelli, Indro, 162
Mont Blanc, 74
Monte dei Paschi di Siena, 194
Montenari, Mr., *108*
Montessori, Elisa, 204–11, *204,* 237
Montezemolo, Giuseppe Cordero di, 117
Montgomery, Bernard L., 101, 115
moon landing, 223–24
Morgan, Ted, 90
Moro, Aldo, 195
Morris, William, 38, 177, 263–65
Mossadegh, Mohammad, 180
MSG 720 B, 247
Mumford, Lewis, 65, 159
Munro, Alice, 230
Murphy, James, 91
Museum of Modern Art, 161, 218
Mussolini, Benito, 44, 47–49, 56, 58, 68, 74, 75, 84, 87, 109, 111, 113–14, 116, 119–21, *120,* 122, 124, 129, 146–48, 150, 244
Adriano Olivetti's plot to remove, 106–9
arrest of, 115, 117
German rescue of, 119–20
Hitler and, 78, 101, 120
Watson and, 99
My Journey at the Nuclear Brink (Perry), 187–88

Nader, Ralph, 239
Naples, 142
NASA (National Air and Space Administration), 223
Nation, 58
National Security Agency (NSA), 235
NATO (North Atlantic Treaty Organization), 149–50, 180, 246, 247, 248
NATO's Secret Armies (Ganser), 147–48, 178–79, 235
Nazi Germany, 63, 75, 78, 84, 94, 95, 98, 101, 104–5, 107, 109, 115, 116, 136, 137, 141, 146, 150, 182, 254
IBM and, 98–99, 116

Nazi Germany *(continued)*
  Italy occupied by, 115–17, 121, 124,
      126, 129–35, 136–37
  Mussolini rescued by, 119–20
NBC, 222
NCR Corporation (National Cash
      Register), 89, 95, 96–97
NEI (New Editions Ivrea), 87, 121
Nesi, Nerio, 244–45, 262–63
News from Nowhere (Morris), 177
Newsweek, 99
New Times, 50
New York, 50–51, 52, 138
  Armory Show in, 63
  World's Fair in,2 20, 221, *222*
New York Times, 138, 222, 255
Nivola, Costantino, *165,* 166
Nizzoli, Marcello, 161, 162
Non Mollare, 68
Normany landings (D-Day), 91, 92, 115
North Afrida, 101, 103–4
nuclear power, 251
nuclear weapons, 149, 179, 186–88, 195,
      208–9, 235
  ICBMs, 208, 234
Nude Descending a Staircase
      (Duchamp), 63
Nutella, 62
Nuzzo, Anna Allegro Olivetti, 79–80
Nuzzo, Antonello, 79

Ochetto, Valerio, 34, 44, 111
O'Hara, Frank, 166
oil industry, 249, 251
Olivetti, 3, 4, 159–64, 228–29, 262
  adding and calculating machines of,
      74, 162, 167, 220–21, 225, 255
  Adriano named director of, 56, 60
  advertisements of, 11, *158*
  Agnelli family and, 61, 242
  American division of, 62, 254
  buildings of, *10,* 11, 36–39, 63–67,
      69–70, *71,* 74, *160,* 174, 175–77,
      256–58, *257,* 261, *261, 263, 264*
  computers of, *see* Olivetti computers
  corporate image of, 86

Cuccia and, 182
design and, 11, 66, 86, 161, 164–65,
      177, 218
Divisumma calculator of, 162, 217
drop in share prices of, 213, 214–15,
      244–45
electronics division of, 184–86,
      188–92, 204, 206, 211, 215–16,
      220, 229, 237, 240–41, 245,
      246, 254, 260; *see also* Olivetti
      computers
factories of, 11, 36–39, *38,* 43, 174,
      175, 261, *261*
foreign companies of, 56
founding of, 27–29, 31
General Electric merger with, 215–16,
      241, 251, 254
growth of, 11, 47, 56, 161, 174
Hispano Olivetti S.A., 56
Ivrea and, 9, 11
Logos 27 calculator of, 220, 221, 225
Massimo as director-general of, 137
Nazis and, 136
Officina Meccanica Olivetti
      (OMO), 47
Olivetti Foundry, 47
Pero named managing director of, 200
political and social philosophy of, 11,
      159
Pozzuoli complex of, 175–77
product lines of, 11
publishing house of, 87, 121, 159
SGS, 190
showrooms of, *163,* 164–67, *165*
Soavi at, 155–57
teletype machines of, 74, 85
typewriters of, *see* Olivetti typewriters
Underwood acquired by, 11–14, 160,
      200–203, *207,* 213, 214, 228, 231,
      237, 244, 246, 254
Underwood headquarters and, 256–57
workers at, 36, 43, 57–58, *57,* 174,
      200
workers housing at, 69–70, *71*
worker takeover idea at, 142
World War II and, 87

Olivetti, Adriano, 3, 10–16, *13*, 17,
    39–42, *41*, 44, 47–59, *48*, 61–62,
    67–68, 70, 78, 84–87, 100–102,
    117–18, 119, 121, 123–25, 131,
    136–41, *140*, 150–54, 155, 159,
    164–69, 173, 186, 196, *202*, 206,
    220, 222, 225, 227, 237, 242, 243,
    254–56, 262–63, 265
  affairs of, 122, 197–99
  American travels of, 50–52, 56, 138,
    200
  arrest and imprisonment of, 118, 119,
    121, 125
  autopsy and, 232–33
  birth of, 30
  break-in at home of, 16, 231, 246
  in Brown affair, 111–14, 117
  childhood of, 31, 33, 34, *34*
  Cola di Rienzo compared to, 62
  Colombo and, 159–61
  Comunità party of, 110–12, 139–40,
    151, 152, 194–96, 200, *202*, 237
  death of, 15, 195, 197, 199, 202, 206,
    208, 224, 228–34, 237, 240, 246,
    251
  Dulles and, 102, 110–11, 194–96, 237
  Elea computer and, 192–94, 204
  electronics and, 185, 204
  European travels of, 54–55, 56
  at factory, 37–39, 44–45
  Fascism and, 58–59, 63, 68
  Ferrarotti and, 137–41
  film about, 37, 230–31
  Ford Motor Company visited by, 52
  funeral for, 15–16, 231, 246
  Giorgio Soavi and, 155–56, 158
  Grazia's marriage to, 172–73, 196–97
  heart attack of, 174
  Heidi and, 197–99
  innovations of, 74, 204, 229
  Ivrea town planning and, 71, 73
  La Piazzola purchased by, 82–84
  memorial to, 262
  military training of, 39–40, 42
  Morris compared with, 264–65
  Mussolini plot of, 106–9

  named director of Olivetti, 56, 60
  Olivetti buildings and, 63–67
  Paola and, 45, 53–54
  Paola's marriage to, 33, 54, 67–68, 74,
    78, 172–73
  Paola's separation from, 78, 79
  personality of, 12–13, 214
  physical appearance of, 12–13, 265
  Pozzuoli complex and, 175–77
  Prandi and, 44
  publishing enterprises of, 87, 151
  psychoanalysis as interest of, 55
  rivalry between Massimo and, 69
  Russia and China and, 204, 237
  Switzerland escape of, 125–27, 128,
    131, 137
  on train before death, 13–15, 143, 231
  Turati's escape and, 49–50, 53, 58,
    69, 101
  Underwood acquired by, 11–14, 160,
    200–203, 213, 214, 228, 231, 237
  Val d'Aosta plans of, 73–74
  Villa Belli Boschi of, 72–73, *73*
  Wanda Soavi and, 117, *117*, 118, 119,
    121–22, 125, 127, 172, 173
Olivetti, Alfred, 154
Olivetti, Anna Allegro, 79–80, 131, 132,
    132, 193
Olivetti, Anna Nogara, 171, 206, 207–8,
    210
Olivetti, Arrigo, 62, 81
Olivetti, Camillo, 9–11, *9*, 17–29,
    30–39, *41*, 42, 47, 53, 60, 62,
    78, 81, *82*, 89, 122–25, 154, 155,
    167–69, 214, 262
  American travels of, 20, 21, 27
  birth of, 17
  Burzio and, 24, 28
  CGS company of, 24, 27, 28
  childhood of, 19, *20*, 30
  convent home of, 31–36, 81, 154,
    260–61
  death of, 125
  education of, 20
  electricity-measuring device of,
    23–24, 27

Olivetti, Camillo *(continued)*
  Fascism and, 59, 63
  as father, 69
  Ferraris and, 20, 21, 23
  funeral for, 125
  household staff and, 37
  marriage of, 26, 27, 30, 37
  money management of, 27
  New Times magazine of, 50
  Olivetti buildings and, 63, 63, 67,
    69–70
  Olivetti workers and, 36, 43, 57–58,
    69–70
  Paola and, 53
  Reform Action paper of, 43–44
  as Socialist, 22, 33, 38–39, 43
  as teenager, *21*
  Turati and, 21–22, 33
  Unitarianism of, 81, 123
  typewriter company founded by,
    27–29, 31
  waterfall memorial to, 17
Olivetti, Camillo (son of Dino and
  Posy), 89
Olivetti, David, 10, *12,* 54, 154, 168,
  173, 224, 241, 242, 252, 253,
  255
Olivetti, Desire, 4, 169–71, 242, 255,
  259
Olivetti, Dino, 10, 13, 14, 30–31, *41,* 54,
  62, 78, 81–82, *87,* 89, 137, 153–54,
  185, 252, 256
  break-in at home of, 16, 231, 246
  in British prison, 88
  marriage of, 82
Olivetti, Elena, 30, 31, *34,* 39, *41,* 54,
  62, 78, 81
Olivetti, Elvira Sacerdoti, 17–19, *18*
Olivetti, Gertrud Kiefer, 10, 81, 85,
  126–28, 153, 154
  escape to Switzerland, 128
Olivetti, Grazia Galletti, 10, 172–74,
  196–99, 232–33, 262
  Adriano's marriage to, 172–73,
    196–97
  break-in at home of, 16, 231, 246

Olivetti, Laura "Lalla" (daughter of
  Adriano), 10, 173, 196–97, 199,
  230, 231, 233
Olivetti, Laura "Lalla" (daughter of
  Camillo), *see* Lizier, Laura "Lalla"
  Olivetti
Olivetti, Lidia, *see* Soavi, Lidia Olivetti
Olivetti, Luisa Revel, 25–27, *25, 26,*
  30–36, *34, 41,* 89, 123–25
  death of, 125
  marriage of, 26, 27, 30, 37
Olivetti, Massimo, 10, 30, 31, *34,* 39,
  41, 62, 67, 74, 78, 81, 85, 126–28,
  137, 153–54
  death of, 154
  marriage of, 85
  as Olivetti director-general, 137
  rivalry between Adriano and, 69
Olivetti, Matteo, 87–88, 241
Olivetti, Paola Levi, 40–42, *45,* 53, *55,*
  67–68, 76, 79, 121, 131–34, 155,
  156, 171, 259
  Adriano and, 45, 53–54
  Adriano's marriage to, 33, 54, 67–68,
    74, 78, 172–73
  Adriano's separation from, 78, 79
  Carlo Levi's portraits of, 76, *76*
  Carlo Levi's relationship with, 77–80
  child born to Carlo and, 79–80
  de Benedetti and, 45–46
  European travels of, 54–55
  La Piazzola villa of, 82, *83,* 129, 133,
    134
  with Roberto in the garden at Fiesole,
    132, 170
Olivetti, Roberto, 3–4, *55,* 67, *67, 83,*
  86, 131, 132, 134, *143,* 167–71, *167,*
  173, *193,* 194, 205–8, *207,* 211,
  212, 214, 215, 222, 224, 226–27,
  226, 229, 241, 242, 244–45,
  251–60, 262
  accidents of, 168
  birth of, 55–56
  computers and, 188, 189
  death of, 3, 259, 260
  electronics laboratory and, 185

Hodgson and, 224, *226*
marriages of, 168, 171, 206, 207, 259
painting celebrating his skill as sailor,
    260
with Paola in the garden at Fiesole,
    132, 170
P101 computer and, *see* Programma
    101 computer
Sottsass's illness and, 192–93
Tchou and, *190,* 191, 206, 260
Tchou's death and, 210–11
Olivetti, Rosamond "Posy" Castle, 14,
    87–88, *87,* 89, 153–54
break-in at home of, 16, 231, 246
Dino's marriage to, 82
Dino's meeting of, 87–88
Olivetti, Rose Emma, 17
Olivetti, Salvador Benedetto, 17–18, *18*
Olivetti, Silvia, 30, 31, 34, *34,* 39, *41,*
    78, 81, 213, 214, 241, 252, 258
Olivetti Builds (Bonifazio and Scrivano),
    29, 73–74
Olivetti computers, 4, 11, 184–86,
    188–94, 229, 240, 246, 248, 251
Elea, 186, 189–91, 193–94, 204, 206,
    209, 238
Programma, *see* Programma 101
    computer
Olivetti Foundation, 263
Olivetti typewriters, 10, 11, 32, 167, 174,
    229, 255
advertisements for, *35*
on assembly line, *203*
Camillo's development of, 27–29, 31
first factory for, 36–39, *38,* 43
keys and touch in, 29
Lettera models, 162, 166
Lexikon 80 model, 161, *161*
M1 model, 29, *29,* 32, 47
M20 model, 47
M40 model, *29,* 56
MP1 model, 56, 162
portable, 56, 74, 162
Studio 42 model, 67
Olmsted, Frederick Law, 20
Orford, Lady, 83

Origo, Iris, 84, 129–30
OSS (Office of Strategic Services), 91,
    92, 95–96, 102, 103, 106, 111, 112,
    118, 147, 238
X-2 branch of, 91, 92, 94, 96
Ottieri, Ottiero, 159
Owen, Robert, 38

P101 (Programma 101 Computer),
    211–26, *212, 219,* 251, 252, 260
Paese Sera, 194
Parmenides, 178, 193
Parri, Ferruccio, 141
Partito Democractico del Livorno, 206
Patton, George S., Jr., 115, 238, 250
Paul VI, Pope, 107
Pavarotti, Luciano, 170
Pavese, Cesare, 137–38
Pearson, Norman Holmes, 91, 92, 93,
    146
Peressutti, Enrico, 70
Pero, Giuseppe, *85,* 87, 136, 137, 200,
    206
death of, 213, 233
Perotti, Luigi, 13, 14, 231
Perotto, Pier Giorgio, 184–86, 188–89,
    206, 211–17, *212,* 220–22, 224,
    225, 228, 245–46, 249, 254
Perrone, Nick, 240
Perry, William J., 187–88, 208, 235
Pershing, John, 89
personality theory, 55
Pescatori, Vittorio, 145
Pétain, Philippe, 101
pharmaceutical companies, 250, 251
Philby, Kim, 89, 94, 144
Philipson, Ruth, 34–35, 53
Phillips, William, 96
Physiology of Love, The (Mantegazza), 39
Piedmont, 119, 139
Pila, 74
Pinsent, Cecil, 84, 133
Pintori, Giovanni, 86
Pirandello, Luigi, 230
Pirelli, 240–41
Pisani, Sallie, 146

Pivato, Marco, 250
Plokhy, Serhii, 232
poison dart guns, 232, *234*
Pollini, Gino, 65, 66, 70, 71, 74, 174
Popular Front, 151, 152
Portrait of Zélide, The (Scott), 84
Pound, Ezra, 90, 91
Powers, Thomas, 92
Pozzuoli, 175–77
Prados, John, 150
Prandi, Giacinto, 44
Prince, The (Machiavelli), 61, 144
Programma 101 computer, 211–26, *212,*
　　219, 251, 252, 260
　called calculator rather than computer,
　　216, 225
　theft of prototype of, 218–20, 231, 246
Proust, Marcel, 46
punch cards, 97–98, 186, 212, 217

Radice, Barbara, 192
RAI, 231
Rao, Giuseppe, 204, 206
Rashomon, 230
Ravizza, Giuseppe, 27–28
Reagan, Ronald, 148
Reform Action, 43–44
Regina Coeli, 116, 118, 119, 125
Reich, Theodore, 138
Remington, 27–28
Revel, Daniel, 25
Revel, Maria, 25–26
Revel, Ulrich, 58
Ribot, 184
Ridenour, Louis, 235
Rocca, Renzo, 244
Rogers, Ernesto Nathan, 70
Romans, ancient, 27, 176
Rommel, Erwin, 101
Roseberry, C. L., 112, 119
Rosselli, Carlo, 75
Rosseli, Nello, 75
Rossi, Ernesto, 68, 75
Rossi, Umberto, 142
Royal Typewriter Company, 51, 202
Ruini, Meuccio, 206

Rumor and Reflection (Berenson), 133
Ruskin, John, 263
Russia, 57, 101, 147, 148–49
　see also Soviet Union

Saarinen, Eero, 221
Sacerdoti, Marco, 18
SAGE, 187
Saibene, Roberto, 256
St. Bernardino convent, 31–36, 81, 154,
　　260–61
Salò Republic, 120, 148
Salvatorelli, Luigi, 107
Salvemini, Gaetano, 44, 48, 58
Sassi community, 175
Scarpa, Carlo, 165
Schawinsky, Alexander "Xanty," 67, 86
Schoenthal, Inge Feltrinelli, 169–70
Scott, Geoffrey, 84, 133
Scrivano, Paolo, 29, 73–74
Secret Surrender, The (Dulles), 104
Secret War in Italy, The (Fowler), 121
Segni, Antonio, 244
Segre, Sion, 69, 75
7-Up, 221
Sforza, Carlo, 111
SGS, 190
Shipman, Tim, 238
Sicily, 114–15, 179
SIFAR, 244
Silicon Valley, 187, 188, 235
Silone, Ignazio, 151
SIM (Servizio Informazioni Militare),
　　118
skyscrapers, 50–51
Smith Corona, 52, 202
Soavi, Albertina, 155–56, *156,* 158
Soavi, Giorgio, 155–59, *157,* 173, 191
Soavi, Lidia Olivetti, *67,* 80, *83,* 131,
　　134, 158–59, *167,* 173
　birth of, 67, 77
　marriage of, 156, 158
Soavi, Michele, 231–32
Soavi, Wanda, 117, *117,* 118, 119, 121–22,
　　125, 127, 155, 172, 173
Social Democrats, 195

Socialism, 22, 33, 38–39, 42–44, 48, 56, 58, 84–85, 138, 151, 152, 180, 182, 194–95, 237, 243–44, 264–65

SOE (Special Operations Executive), 111–14, 117–18, 119

Sottsass, Ettore, 191–93, *193,* 206, 217, 256
  illness of, 192–93

Sovereign Military Order of Malta (SMOM), 147

Soviet Union, 59, 148–49, 152, 178–80, 187, 188, 195, 204, 208, 232, 235–38, 240, 246–49
  Italy and, 243

Spain, 63, 98

Spanzotti, Giovanni Martino, 31–32

Stabler, Wells, 180

Stalin, Joseph, 148, 178

Stampa, 107, 209, 210, 261

Standard Oil, 98

State Department, 152, 243

Stein, Gertrude, 83

Steiner, Rudolf, 39

Sutherland, Graham, 157, *157*

Switzerland, 109, 110, 125–28, 131, 133, 137
  smuggling between Italy and, 218

Tagliabue, John, 255

Talbot, David, 103, 105, 106, 243

Talucchi, Giuseppe Maria, 72

Tambroni, Fernando, 195

Taylor, Maxwell D., *108*

Tchou, Elisa Montessori, 204–11, *204,* 237

Tchou, Mario, 185, 188–91, 194, 204–8, 222, 224, 237, 254
  China trip of, 206–8, 237, 246
  death of, 208–11, *210,* 228–29, *234,* 237–40, 251
  marriage of, 204–5
  Roberto Olivetti and, *190,* 191, 206, 260

Tchou, Nicola, 210, 211

Tchou Yin, 185

teletype machines, 74, 85, 237

Tempest program, 236, 247, 248

39 Steps, The, 232

Tinesi, Carlo, 210

Tobino, Mario, 79

Togliatti, Palmiro, 151

Toklas, Alice B., 83

Tolstoy, Leo, 263

Toppi, Giancarlo, *212*

Tower, John G., *234*

Truman, Harry, 152

Turati, Filippo, 21–22, *22,* 33
  house arrest and escape of, 49–50, 53, 58, 69, 101

Turin, 18, 43, 75, 88, 101, 136

typewriters, 174, 202
  electric, 28
  Olivetti, *see* Olivetti typewriters
  portable, 52, 74
  Ravizza, 27–28
  Remington, 27–28
  Royal, 51, 202
  Smith Corona, 52, 202
  Underwood, *see* Underwood

Umberto II of Italy, 106, *106,* 107, 109, 151

Underwood, 11–12, 22, 47, 51, 201–3, 213
  headquarters of, 256–57
  Olivetti's acquisition of, 11–14, 160, 200–203, *207,* 213, 214, 228, 231, 237, 244, 246, 254

Unità, 215

Unitarianism, 81, 123

United Nations, 175

United Nations Educational, Scientific, and Cultural Organization (UNESCO), 263

United States:
  Adriano Olivetti in, 50–52, 56, 138, 200
  Air Force of, 187, 224, 235
  Camillo Olivetti in, 20, 21, 27
  China and, 208
  CIA in, *see* CIA
  isolationism in, 98

United States *(continued)*
  Italy and, 151–52, 178, 180, 181, 243, 243, 248–51
  Olivetti company in, 62, 254
  Olivetti-Underwood merger and, 202
  OSS in, *see* OSS
  State Department of, 152, 243
  in World War I, 98
  World War II entry of, 84
Universal Exhibition, 32
University of Chicago, 139
Unruly, 145
U.S. Steel, 98

Val d'Aosta, 73–74
Valente, Maria, 247–48
Vallesa, Alessandro, 72
Valletta, Vittorio, 142, 181, *183,* 216, 241, 246, 253
Vanderbilt, William, 95–96
Vanity Fair, 198
Via Tasso, 116–17
Victor Emanuel III of Italy, 108–9, *109*
Vidal, Gore, 162
Vietnam War, 224
Villa, Pancho, 89
Villa Belli Boschi, 72–73, *73*
Villa Casana, 72
Villa i Tatti, 83, 84, 132–33
Villa Medici, 83, 129, 133
Villa Savoie, 64, 65, *65*
Visconti, Luchino, 162
Visentini, Bruno, 214, *214,* 220, 225, 241, 252–54
Vitale, Alberto, 162, 164
Volponi, Paolo, 159

Waldensians, 25, 78
Waldo, Peter, 25
Waller, Douglas, 93
Wall Street Journal, 222
Walpole, Horace, 83
Walters, Vernon, 243
War in Italy (Lamb), 115

Warren Commission, 233
Watson, Thomas J., 96–98
  Hitler and, 99
  Mussolini and, 99
Weiner, Tim, 149
Weller, Charles, 166
Westinghouse, 21
White, E. B., 50–51
Whittle, David W., 223–24
William Morris: A Life for Our Times (MacCarthy), 265
Wilson, Woodrow, 98
Windsor, Duke and Duchess of, 99
Winks, Robin, 91, 92, 94, 95
work environment studies, 85
World's Columbian Exposition, 20–21
World War I, 42–43, 44, 95, 109–10, 111
  Dulles in, 103
  U.S. in, 98
World War II, 78, 81, 84, 85, 87–89, 95, 101–5, 107, 137, 149, 186
  D-Day in, 91, 92, 115
  end of, 136, 148, 188
  espionage and counterintelligence during, 89–94, 103
  London bombed during, 93
  Olivetti company and, 87
  Sicily invasion in, 114–15, 179
  U.S. entry into, 84
  see also Nazi Germany
Wright, Frank Lloyd, 27, 38, 74, 174, 256, 257–58
  Johnson and, 64–65

X-2, 91, 92, 94, 96

Yes, Prime Minister, 113

Zamara, Jose, 239
Zevi, Bruno, 198
Zingaretti, Luca, 231
Zorzi, Renzo, 195–96
Zuccotti, Susan, 116

Illustration Permissions

My thanks go to David and Philip Olivetti, sons of Dino Olivetti, who first encouraged me to undertake this project and helped me at every turn. David Olivetti, in particular, has assembled a priceless collection of photographs of the family going back for many decades. Many of them are contained in his major oeuvre, Album Di Famiglia Olivetti, giving me full and generous access to the treasures within. The same is true of photographs in the files at the Fondazione Adriano Olivetti in Rome, under the kind permission of its CEO, Beniamino de' Liguori Carino, and the Archivio Storico Olivetti in Ivrea.

Courtesy of the Fondazione Adriano Olivetti: 9, 13, 18 (2), 20, 21, 23, 25, 26, 34, 35, 38, 41, 45, 48, 55, 57, 67, 71, 82, 85, 87, 140, 143, 158, 160, 161, 163, 165, 190, 202, 203, 207, 212, 204, 219, 257

Courtesy of the author: 6, 7, 8, 10, 12, 65, 73, 78, 90, 92, 102, 106, 108, 109, 114, 117, 120, 179, 210, 222, 234, 261, 264

Courtesy of Elisa Montessori: 204

Courtesy of Milton Gendel: 205

Courtesy of Anna Olivetti: 76, 83, 260

Courtesy of Albertina Soavi: 156, 157, 132, 167, 190, 193, 226

Courtesy of Anna Nogara & Desire Olivetti: 132, 167, 190, 193, 226

Courtesy of Roberto Battegazzorre: 183, 220

Alamy Stock Photo: 22, 48, 182, 214, 247, 250

A Note on the Type

This book was set in Adobe Garamond. Designed for the Adobe Corporation by Robert Slimbach, the fonts are based on types first cut by Claude Garamond (c. 1480–1561). Garamond was a pupil of Geoffroy Tory and is believed to have followed the Venetian models, although he introduced a number of important differences, and it is to him that we owe the letter we now know as "old style."

Composed by North Market Street Graphics
Lancaster, Pennsylvania

Printed and bound by Berryville Graphics
Berryville, Virginia

Designed by Michael Collica